Basic Biomechanics
of the
Skeletal System

Basic Biomechanics
of the
Skeletal System

VICTOR H. FRANKEL, M.D., DR. SCI.

Director, Orthopaedic Surgery
Orthopaedic Institute
Hospital for Joint Diseases
New York, New York

Professor of Orthopaedic Surgery
Mount Sinai School of Medicine of the City of New York
New York, New York

MARGARETA NORDIN, R.P.T.

Associate in the Department of Orthopaedic Surgery
Sahlgren Hospital
University of Göteborg, Göteborg, Sweden

and

Associate in Orthopaedics
Orthopaedic Institute
Hospital for Joint Diseases
New York, New York

KAJSA FORSSÉN, *Illustrator*
LAURIE GLASS, *Consulting Editor*

 LEA & FEBIGER   *Philadelphia*

Library of Congress Cataloging in Publication Data

Frankel, Victor H
　　Basic biomechanics of the skeletal system.

　　Includes bibliographies and index.
　　1.　Human mechanics.　　I.　Nordin, Margareta, joint
author.　　II.　Title.
QP303.F66　　　　　　612'.75　　　　　　79-24593
ISBN　0-8121-0708-X

PRINTED IN THE UNITED STATES OF AMERICA

Print Number　3　2

*The authors wish to dedicate this book
to the memory of Professor Carl Hirsch,
the founder of modern biomechanics.*

Foreword

Progress in restoring function has paralleled our gain in knowledge of the musculoskeletal system and the nature of its disabilities. Observation, thought and experience were sufficient to solve some problems. Further answers evolved as we became more cognizant of underlying pathology and physiology. Many of the remaining problems await the application of biomechanical principles.

A major determinant of effective function is the force tolerance and mobility of the skeletal system. Increases in the intensity of activity of all age groups and extended human longevity are presenting new demands on these physical qualities. To develop an appropriate clinical response biomechanical information must be acquired and the knowledge translated into therapeutic guidelines.

Victor Frankel has been a leader in both these endeavors. His contribution to biomechanical research is attested to by his extensive publications. This text represents his second effort to introduce biomechanical knowledge into patient care. It is not a simple task. Margareta Nordin has joined him in this responsibility for producing a new text. Her involvement has contributed to the informative style used to interpret complex data. The value of this text also has been enhanced by the added breadth provided by the contributors.

The engineering profession has developed the means for identifying the forces, motions, and tissue responses in human function. A considerable amount of information has been provided, but this information is dispersed in a wide variety of publications and is often couched in engineering terminology unfamiliar to clinicians.

This book on the basic biomechanics of the skeletal system brings together the current knowledge and presents it in a clear, conversational style. The use of engineering terminology is restricted to that which is necessary to express essential concepts. Presentation of the literature is both inclusive and selective. The major contributions to each topic are cited as specific references and other sources are identified as recommended reading.

While the text is primarily a compendium of information and illustrations of biomechanical principles which are clinically pertinent, a small portion of the book describes the means for calculating joint and muscle forces. This is presented lucidly with minimal mathematical formulae and very clear diagrams. The techniques are described in a way which makes them usable by the interested clinician.

This book is a well conceived bridge between engineering concepts and clinical practice. It should provide orthopaedists, physical and occupational therapists, and other health professionals with a working knowledge of biomechanical principles useful in the evaluation and treatment of musculoskeletal dysfunction.

Jacquelin Perry, M.D.

Preface

This book has been written to serve as an introduction to biomechanics for those who deal with disorders of the skeletal system. This work is the result of 20 years of experience in biomechanics education and many conferences with colleagues in the fields of orthopaedic surgery, physical and occupational therapy, and other allied specialties.

Biomechanics uses laws of physics and engineering concepts to describe motion undergone by the various body segments and the forces acting on these body parts during normal daily activities. The interrelationship of force and motion is important, and must be understood if rational treatment programs are to be applied to skeletal disorders. Deleterious effects may be produced if during exercise or activity the forces acting on disordered parts rise to high levels.

The purpose of this volume is to acquaint the readers with the forces and motions acting in the skeletal system and the various techniques used to elucidate them. It is intended to be used as a textbook, in conjunction with a biomechanics course, or for independent study; the references and suggested readings at the end of each chapter may be utilized to amplify the discussions in these chapters. Obviously, this text can only serve as a guide to a deeper understanding of skeletal biomechanics through further reading, analysis of case material, and independent research. No attempt has been made to discuss therapy, since it is not the purpose of the text to cover this area. Rather, we have described the underlying basis for a rational therapeutic program.

This book has been strengthened by the work of the contributing authors. Dennis Carter's chapter on the international system of measurements serves as an introduction to the important physical measurements used throughout the book. It is not necessary to have more than a basic knowledge of mathematics to fully comprehend the material in the book, but it is important to first begin by reviewing this section on the SI system and its application to biomechanics. Van Mow, Vladimir Roth, and Cecil Armstrong have elucidated important concepts in joint lubrication in their chapter on biomechanics of joint cartilage. James Sammarco has synthesized a great deal of material on biomechanics of the foot. Frederick Matsen has contributed chapters on the shoulder and elbow joint which serve as an up-to-date review of these somewhat underemphasized areas. Margareta

Lindh has summarized the vast area of biomechanics of the lumbar spine and has covered the application of this area to clinical situations.

In two chapters, basic principles have been developed that are expanded upon in subsequent chapters. Chapter 1, Biomechanics of Whole Bones and Bone Tissue, discusses the mechanical properties of materials. Chapter 4, Biomechanics of the Knee, illustrates important concepts of kinematics and force analysis. It is suggested that these chapters be studied first.

Although we undertook an extensive review of the world literature on biomechanics in preparing to write this book, it has not been our purpose to publish a review of the material, but rather to select examples to illustrate the concepts necessary for a knowledge of skeletal biomechanics. Important engineering concepts have been developed throughout the volume.

The material presented in this volume should act as a guide to the reader in the assessment of the literature on biomechanics. We feel that our efforts in writing this book will be justified if it brings about an increased awareness of the importance of biomechanics and engenders discussion.

Seattle, Washington   Victor H. Frankel
Göteborg, Sweden   Margareta Nordin

Acknowledgments

This book was made possible through the intense editorial efforts of Laurie Glass. Not only did she function as grammarian and stylist, but her logical pattern of thinking had a great deal to do with the final organization of the book.

Kajsa Forssén, illustrator, was the fourth member of our publication team. Her quick grasp of the concepts to be illustrated and the clarity of her art work were of the greatest importance in producing this book.

Constructive criticism of various chapters was given by Professor Alf Nachemson and Drs. Gunnar Andersson, Dennis Carter, Theodore Greenlee, Göran Lundborg, and Mary Ann Riederer-Henderson.

No book is created without a capable and energetic production corps. Ours was the best. Bea Watts devoted an extraordinary amount of time and effort to this volume and has our deep gratitude. A special thanks is extended to Sandra Greenlee for her support and help. Margit Hobbs, Sarah Sato, Pat Schuppenhauer, and Linda Sutch also devoted their time to manuscript preparation. We are grateful to Susan Russell for capably handling the final stages of figure production.

This work was undertaken while Margareta Nordin was a Rotary International Foundation fellow in the Department of Orthopaedics at the University of Washington. Support was also received from the Reene Eanders hjälpfond and the Legitimerade Sjukgymnaster Riksförbund. The generosity of these organizations made this international cooperation possible.

To all who helped, hjärtligt tack!

Victor H. Frankel
Margareta Nordin

Contributors

Cecil G. Armstrong, Ph.D.
Research Associate
Department of Mechanical Engineering,
Aeronautical Engineering and Mechanics
Rensselaer Polytechnic Institute
Troy, New York

Dennis R. Carter, Ph.D.
Director of Orthopaedic Biomechanics
Massachusetts General Hospital
Assistant Professor of Orthopaedic Surgery
Harvard Medical School
Boston, Massachusetts

Victor H. Frankel, M.D., Dr. Sci.
Director, Orthopaedic Surgery
Orthopaedic Institute
Hospital for Joint Diseases
New York, New York

Professor of Orthopaedic Surgery
Mount Sinai School of Medicine of the City of New York
New York, New York

Margareta Lindh, R.P.T., Ph.D.
Department of Orthopaedic Surgery
Sahlgren Hospital
University of Göteborg
Göteborg, Sweden

Frederick A. Matsen III, M.D.
Associate Professor
Department of Orthopaedics
University of Washington
Seattle, Washington

Van C. Mow, Ph.D.
Professor of Mechanics and Biomedical Engineering
Department of Mechanical Engineering,
Aeronautical Engineering and Mechanics
Rensselaer Polytechnic Institute
Troy, New York

Margareta Nordin, R.P.T.
Associate in the Department of Orthopaedic Surgery
Sahlgren Hospital
University of Göteborg
Göteborg, Sweden
   and
Associate in Orthopaedics
Orthopaedic Institute
Hospital for Joint Diseases
New York, New York

Vladimir Roth, Ph.D.
Research Associate
Department of Mechanical Engineering,
Aeronautical Engineering and Mechanics
Rensselaer Polytechnic Institute
Troy, New York

G. James Sammarco, M.D., F.A.C.S.
Associate Clinical Professor
Department of Orthopaedics
University of Cincinnati Medical Center
Cincinnati, Ohio

Contents

SI: THE INTERNATIONAL SYSTEM OF UNITS

Dennis R. Carter

The need for establishing systems of weights and measures for use in building, making clothes, and engaging in simple trade and commerce was recognized by primitive societies. The Bible and early records from Babylonian and Egyptian civilizations indicate that man first used parts of his body and common elements in his environment to establish simple systems of measurement. Time was commonly measured by periods of the sun. At the time of Noah, length was measured in terms of the cubit, which was equivalent to the distance from the elbow to the tip of the middle finger. A span, equivalent to one-half a cubit, was taken as the distance from the tip of the thumb to the tip of the little finger with the hand fully spread. For smaller measurements the digit, or width of the thumb, was used. To measure volumes or fluid capacities, containers were filled with seeds, which were then poured out and counted to establish the volume of the container in terms of the number of seeds it could hold.

Primitive societies found that they could determine if one object was heavier than another by placing the two objects on a balance. The Babylonians refined this technique by balancing an object with a set of standard, well-polished stones. These stones were probably the first standards, and the measurement later evolved into the English legal stone, which in the imperial system is equivalent to 14 pounds (62.27 newtons).

The Egyptians and the Greeks both used the wheat seed as the standard for the smallest unit of weight. This concept evolved into the unit of the grain, which is still used today in limited applications. The Arabs established a standard of weights for precious stones and metals based on the weight of a small bean called a Karob, a measurement which has evolved into the present unit of the carat.

Increasing trade among tribes and nations caused the measuring systems developed by early civilizations to become intermixed. Measuring standards were widely disseminated by the Romans throughout the world. As the Roman soldiers marched in their conquests, distances were measured in terms of the pace, which was the distance between the points where one foot struck the ground on successive steps. A pace was therefore equivalent to two steps.

THE ENGLISH SYSTEM

The English system of weights and measures evolved from measuring systems used by the Babylonians, Egyptians, Romans, Anglo-Saxons, and

Norman French. Body measurements such as the digit, palm, span, and cubit were replaced by the inch, foot, and yard. The Romans contributed the use of the base unit 12 to the English system, as the Roman foot, *pes,* had 12 divisions called *unciae.* The English words *inch* and *ounce* were derived from this Latin word.

The early Saxon kings wore a sash or girdle around their waists which could be removed conveniently and used for linear measurement. The word *yard* was derived from the Saxon word *gird,* meaning the circumference of a person's waist. In the twelfth century, King Henry I decreed the standard yard to be equivalent to the distance between his nose and the end of his thumb with his arm extended. In the thirteenth century, King Edward I took an important step forward in standardizing linear measurement by ordering a permanent measuring stick made of iron to be used as the standard yard for the kingdom. This measurement stick was coined the "iron ulna" in reference to the bone of the forearm. The measurement of a foot was established as one-third the length of the measuring stick, and an inch was established as one thirty-sixth of the standard yard. King Edward II apparently found this new measuring system to be confusing and passed a statute decreeing an inch to be equivalent to "three barleycorns, round and dry."

The English system of measurement evolved and became refined primarily through royal decrees. The early Tudor rulers established 220 yards as equivalent to 1 furlong. In the sixteenth century, Queen Elizabeth I changed the traditional Roman mile from 5,000 to 5,280 feet, making the mile equivalent to exactly 8 furlongs. In 1824 the English Parliament legalized the new standard yard, which was based on the measurements taken from a brass bar with a gold button near each end. A single dot was engraved on each button, and the yard was taken to be the distance between the two engraved dots.

The English colonization and dominance of world commerce in the seventeenth, eighteenth, and nineteenth centuries served to spread the English system of weights and measures throughout the world, including the American colonies. Although refined in some ways to meet the demands of commerce, the system has remained essentially unchanged.

THE METRIC SYSTEM

In the early eighteenth century no uniform system of weights and measures existed on the European continent. Measurements not only differed from country to country, but also from town to town. During the French Revolution in 1790, King Louis XVI and the National Assembly of France passed a decree calling on the French Academy of Sciences and the Royal Society of London to "deduce an invariable standard for all of the measures and all weights." The English, however, did not participate in this undertaking, and so the French alone set out to establish a new and uniform system of measures. The result was the metric system.

In 1793 the French government adopted a standard system of measurement based on the meter, which was defined as one ten-millionth of the distance from the North Pole to the equator on a line passing through Paris. All other linear measurements were established as decimal fractions of the meter. The metric unit of mass was established as the gram, which was defined as the mass of 1 cubic centimeter of water at its temperature of maximum density. The unit of fluid capacity was named the "litre" and defined as the volume contained by a cube which measured one-tenth of a meter on each side.

Most Frenchmen thought that the new metric system of measurement was confusing and expended a great deal of effort converting from the new system to the familiar measuring system of yards and feet that had been used previously. Widespread resistance to the new system forced Napoleon to renounce the metric system in 1812. In 1837, however, France returned to the metric system in the hope that it would spread throughout the world.

In the nineteenth century, the metric system found favor among scientists because (1) it was an international system, (2) the units were independently reproducible, and (3) its use of a decimal system made calculations much simpler. The British system had been designed for trade and commerce, but the metric system appeared to have significant advantages in the areas of engineering and science.

In the latter half of the nineteenth century, scientific advances necessitated the development of better metric standards. In 1875 an international treaty, the Treaty of the Meter, established well-defined metric standards for length and mass. In addition, the permanent machinery for further refinements of the metric system was established. This treaty was signed by 17 countries. By 1900, 35 nations had officially adopted the metric system.

At the time the Treaty of the Meter was signed, a permanent secretariat, the International Bureau of Weights and Measures, was established in Serves, France, to coordinate the exchange of information about the metric system. The diplomatic organization concerned with the metric system is the General Conference of Weights and Measures, which meets periodically to ratify improvements in the system and the standards. In 1960, the general conference adopted an extensive revision and simplification of the system. This modernized metric system was named Le Systeme International d'Unites (International System of Units), and was given the international abbreviation SI. The general conference adopted improvements and additions to the SI system in 1964, 1968, 1971, and 1975.

Between 1965 and 1972, the United Kingdom, Australia, Canada, New Zealand, and many other English-speaking countries adopted the metric system. In 1975, the United States Congress passed the Metric Conversion Act, providing for the adoption of the SI metric system as the predominant system of measurement units. Today, nearly the entire world is either using the metric system or committed to its adoption.

THE SI METRIC SYSTEM

The Systeme International d'Unites (SI) is the modern metric system which has evolved into the most exacting system of measures devised. In this section, the SI units of measurement used in the science of mechanics are described. SI units used in electrical and light sciences have been omitted for the sake of simplicity.

The SI units can be considered in three groups: (1) the base units, (2) the supplementary units, and (3) the derived units (Fig. 1). The base units are a small group of standard measurements which have been arbitrarily defined. The base unit for length is the meter (m), and the base unit of mass is the kilogram (kg). The base units for time and temperature are the second (s) and the kelvin (K), respectively. Definitions of the base units have become increasingly sophisticated in response to the expanding needs and capabilities of the scientific community (Table 1). For example, the meter is now defined in terms of the wavelength of radiation emitted from the krypton-86 atom.

The radian (rad) is a supplementary unit to measure plane angles. This unit, like the base units, is arbitrarily defined (Table 1). Although the radian is the SI unit for plane angle, the unit of the degree has been retained for

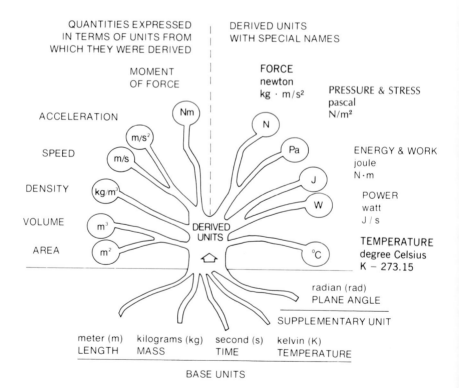

Fig. 1. The International System of Units.

TABLE 1   DEFINITIONS OF SI UNITS

BASE SI UNITS

meter (m)	The meter is the length equal to 1,650,763.73 wavelengths in vacuum of the radiation corresponding to the transition between the levels $2p_{10}$ and $5d_5$ of the krypton-86 atom.
kilogram (kg)	The kilogram is the unit of mass and is equal to the mass of the international prototype of the kilogram.
second (s)	The second is the duration of 9,192,631,770 periods of the radiation corresponding to the transition between the two hyperfine levels of the ground state of the cesium-133 atom.
kelvin (K)	The kelvin, a unit of thermodynamic temperature, is the fraction 1/273.16 of the thermodynamic temperature of the triple point of water.

SUPPLEMENTARY SI UNIT

radian (rad)	The radian is the plane angle between two radii of a circle which cut off on the circumference an arc equal in length to the radius.

DERIVED SI UNITS WITH SPECIAL NAMES

newton (N)	The newton is that force which, applied to a mass of one kilogram, gives it an acceleration of one meter per second squared. 1 N = 1 kg m/s².
pascal (Pa)	The pascal is the pressure produced by a force of one newton applied, uniformly distributed, over an area of one square meter. 1 Pa = 1 N/m².
joule (J)	The joule is the work done when the point of application of a force of one newton is displaced through a distance of one meter in the direction of the force. 1 J = 1 Nm.
watt (W)	The watt is the power which in one second gives rise to energy of one joule. 1 W = 1 J/s.
degree Celsius (°C)	The degree Celsius is a unit of thermodynamic temperature and is equivalent to K − 273.15.

general use, since it is firmly established and widely used around the world. A degree is equivalent to $\pi/180$ rad.

Most units of the SI system are derived units, meaning that they are established from the base units in accordance with fundamental physical principles. Some of these units are expressed in terms of the base units from which they are derived. Examples are area, speed, and acceleration, which are expressed in the SI units of square meters (m²), meters per second (m/s), and meters per second squared (m/s²), respectively.

Other derived units are similarly established from the base units but have been given special names (see Fig. 1 and Table 1). These units are defined through the use of fundamental equations of physical laws in conjunction with the arbitrarily defined SI base units. For example, Newton's second law of motion states that when a body which is free to move is subjected to a force, it will experience an acceleration proportional to that force and inversely proportional to its own mass. Mathematically this principle can be expressed as

$$\text{force} = \text{mass} \times \text{acceleration}.$$

The SI unit of force, the newton (N), is therefore defined in terms of the base SI units as

$$1 \text{ N} = 1 \text{ kg} \times 1 \text{ m/s}^2.$$

The SI unit of pressure and stress is the pascal (Pa). Pressure is defined in hydrostatics as the force divided by the area of force application. Mathematically this can be expressed as

$$\text{pressure} = \frac{\text{force}}{\text{area}}.$$

The SI unit of pressure, the pascal (Pa), is therefore defined in terms of the base SI units as

$$1 \text{ Pa} = \frac{1 \text{N}}{1 \text{m}^2}.$$

Although the SI base unit of temperature is the kelvin, the derived unit of degree Celsius (°C or c) is much more commonly used. The degree Celsius is equivalent to the kelvin in magnitude, but the absolute value of the Celsius scale differs from that of the Kelvin scale such that °C = K − 273.15.

When the SI system is used in a wide variety of measurements, the quantities expressed in terms of the base, supplemental, or derived units may be either very large or very small. For example, the area on the head of a pin is an extremely small number when expressed in terms of square meters (m^2). On the other hand, the weight of a whale is an extremely large number when expressed in terms of newtons (N). To accommodate the convenient representation of small or large quantities, a system of prefixes has been incorporated into the SI system (Table 2). Each prefix has a fixed meaning and can be used with all SI units. When used with the name of the unit, the prefix indicates that the quantity described is being expressed in some multiple of ten times the unit used. For example, the millimeter (mm) is used to represent one-thousandth (10^{-3}) of a meter and a gigapascal (GPa) is used to denote one billion (10^9) pascals.

One of the more interesting aspects of the SI system is its use of the names of famous scientists as standard units. In each case, the unit was named after a scientist in recognition of his contribution to the field in which that unit plays a major role. Table 3 lists a number of SI units and the scientist for which each was named.

The unit of force, the newton, was named in honor of the English scientist Sir Isaac Newton (1624–1727). He was educated at Trinity College at Cambridge and later returned to Trinity College as a professor of mathematics. Early in his career he made fundamental contributions to mathematics which formed the basis of differential and integral calculus. His other major discoveries were in the fields of optics, astronomy, gravitation, and mechan-

TABLE 2   SI PREFIXES

Prefix	Symbol	Meaning	Exponent or Power
exa	E	quintillion	10^{18}
peta	P	quadrillion	10^{15}
tera	T	trillion	10^{12}
giga	G	billion	10^{9}
mega	M	million	10^{6}
kilo	k	thousand	10^{3}
hecto	h	hundred	10^{2}
deka	da	ten	10^{1}
		base unit	
deci	d	tenth	10^{-1}
centi	c	hundredth	10^{-2}
milli	m	thousandth	10^{-3}
micro	μ	millionth	10^{-6}
nano	n	billionth	10^{-9}
pico	p	trillionth	10^{-12}
femto	f	quadrillionth	10^{-15}
atto	a	quintrillionth	10^{-18}

ics. His work in gravitation was purportedly spurred by being hit on the head by an apple falling from a tree. It is perhaps poetic justice that the SI unit of one newton is approximately equivalent to the weight of a medium-sized apple. Newton was knighted in 1705 by Queen Mary for his monumental contributions to science.

The unit of pressure and stress, the pascal, was named after the French physicist, mathematician, and philosopher Blaise Pascal (1623–1662). Pascal conducted important investigations on the characteristics of vacuums and barometers and also invented a machine which would make mathematical calculations. His work in the area of hydrostatics and hydrodynamics helped lay the foundation for the later development of these scientific fields. In addition to his scientific pursuits, Pascal was passionately interested in religion and philosophy and thus wrote extensively on a wide range of subjects.

The base unit of temperature, the kelvin, was named in honor of Lord William Thomson Kelvin (1824–1907). He was of Scottish-Irish descent, and his given name was William Thomson. He was educated at the University of Glasgow and Cambridge University and early in his career investigated the thermal properties of steam at a scientific laboratory in Paris.

TABLE 3 SI UNITS NAMED AFTER SCIENTISTS

Symbol	Unit	Quantity	Scientist	Country of Birth	Dates
A	ampere	electric current	Ampere, Andre-Marie	France	1775–1836
C	coulomb	electric charge	Coulomb, Charles Augustin de	France	1736–1806
°C	degree Celsius	temperature	Celsius, Anders	Sweden	1701–1744
F	farad	electric capacity	Faraday, Michael	England	1791–1867
H	henry	inductive resistance	Henry, Joseph	United States	1797–1878
Hz	hertz	frequency	Hertz, Heinrich Rudolph	Germany	1857–1894
J	joule	energy	Joule, James Prescott	England	1818–1889
K	kelvin	temperature	William Thomson, Lord Kelvin	England	1824–1907
N	newton	force	Newton, Sir Isaac	England	1642–1727
Ω	ohm	electric resistance	Ohm, Georg Simon	Germany	1787–1854
Pa	pascal	pressure/stress	Pascal, Blaise	France	1623–1662
S	siemens	electric conductance	Siemens, Karl Wilhelm later Sir William	Germany (England)	1823–1883
T	tesla	magnetic flux density	Tesla, Nikola	Croatia (United States)	1856–1943
V	volt	electrical potential	Volta, Count Alessandro	Italy	1745–1827
W	watt	power	Watt, James	Scotland	1736–1819
Wb	weber	magnetic flux	Weber, Wilhelm Eduard	Germany	1804–1891

At the age of 32 he returned to the University of Glasgow to accept the chair of Natural Philosophy. His meeting with James Joule in 1847 stimulated interesting discussions on the nature of heat which eventually led to the establishment of Thomson's absolute scale of temperature, the Kelvin scale. In recognition of Thomson's contributions to the field of thermodynamics, King Edward VII conferred on him the title of Lord Kelvin.

The commonly used unit of temperature, the degree Celsius, was named after the Swedish astronomer and inventor Anders Celsius (1701–1744). Celsius was appointed professor of astronomy at the University of Uppsala at the age of 29 and remained at the university until his death 14 years later. In 1742 he described the centigrade thermometer in a paper prepared for the Swedish Academy of Sciences. The name of the centigrade temperature scale was officially changed to Celsius in 1948.

CONVERTING TO SI FROM OTHER UNITS OF MEASUREMENT

Table 4 has been provided to allow conversion of measurements expressed in English and non-SI metric units into SI units. One fundamental source of confusion in converting from one system to another is that two basic types of measurement systems exist. In the "physical" system (such as SI) the units of length, time, and *mass* are arbitrarily defined, and other units (including force) are derived from these base units. In "technical" or "gravitational" systems (such as the English system) the units of length, time, and *force* are arbitrarily defined, and other units (including mass) are derived from these base units. Since the units of force in gravitational systems are in fact the *weights* of standard masses, conversion to SI is dependent upon the

TABLE 4   CONVERTING TO SI FROM OTHER UNITS

Length				
To convert to meters (m)				
	multiply	in	by	0.0254
		ft		0.3048
		yd		0.9144
		mile		1609.3
Area				
To convert to square meters (m²)				
	multiply	in^2	by	6.4516×10^{-4}
		ft^2		0.0929
		yd^2		0.8361
Volume				
To convert to cubic meters (m³)				
	multiply	in^3	by	1.6387×10^{-5}
		ft^3		0.0283
		yd^3		0.7646

(Note: A liter is equivalent to $1 \text{ m}^3 \times 10^{-3}$.)

TABLE 4   (Continued)

Mass
To convert to kilograms (kg)

	multiply	lb	by	0.4536
		slug		14.594
		kgf s^2 m^{-1}		9.8067

Density
To convert to kg/m^3

	multiply	lb/in^3	by	27,680
		lb/ft^3		16.018

Moment of Inertia
To convert to kg m^2 or, equivalently, to Nm sec^2

	multiply	lb ft^2	by	0.0421
		lb in^2		2.9264 $\times$ 10^{-4}
		kgf s^2 m		9.8067

Force
To convert to newtons (N)

	multiply	kgf (=kp)	by	9.8067
		lbf		4.4482

Pressure and Stress
To convert to pascals (Pa) or, equivalently, to N/m^2

	multiply	lbf/in^2	by	6894.8
		kgf/m^2		9.8067
		mm Hg		133.32

Energy, Work, and Moment of Force
To convert to joules (J) or, equivalently, to Nm

	multiply	ft lbf	by	1.3558
		in lbf		0.1130
		kgf m		9.8067
		Btu		1055.1
		cal		4.1868

Time
To convert to seconds (s)

	multiply	min	by	60
		h		3600
		d		86,400

Speed
To convert to m/s

	multiply	ft/s	by	0.3048
		km /h		0.2778
		mile/h		0.4470

Acceleration
To convert to m/s^2

	multiply	ft/s^2	by	0.3048

Temperature
To convert to degrees Celsius

from deg K: deg C = deg K − 273.15
from deg F: deg C = 5/9 (deg F − 32)

Plane Angle
To convert to radians (rad)

	multlipy	deg	by	π/180 (or 0.01745)
		rev		2π (or 6.2832)

acceleration of mass due to the Earth's gravity. By international agreement the acceleration due to gravity is 9.806650 m/s². This value has been used in establishing some of the conversion factors in Table 4.

BIBLIOGRAPHY

Feirer, J. L.: *SI Metric Handbook.* New York, Charles Scribner's Sons, 1977.
Pennycuick, C. J.: *Handy Matrices of Unit Conversion Factors for Biology and Mechanics.* New York, John Wiley and Sons, 1974.
World Health Organization. *The SI for the Health Professions.* Geneva, World Health Organization, 1977.

Part One

Biomechanics of Tissues and Structures of the Skeletal System

CHAPTER 1

Biomechanics of Whole Bones and Bone Tissue

Margareta Nordin
and
Victor H. Frankel

The purpose of the skeletal system is to protect internal organs, provide rigid kinematic links and attachment sites for muscles, and facilitate muscle action and body movement. Bone has unique mechanical properties that allow it to carry out these roles.

Bone is self-repairing and can alter its properties and configuration in response to changes in mechanical demand. For example, changes in bone density are commonly observed after periods of either disuse or greatly increased function. Changes in bone shape are observed during fracture healing and after certain operative procedures. Thus, bone appears to adapt to the mechanical demands placed upon it.

This chapter will describe the mechanical properties of normal whole bones and bone tissue, the various loading modes to which bone is subjected, and the factors that affect bone strength and stiffness.

MECHANICAL PROPERTIES OF BONE

Strength and stiffness are the important mechanical properties of bone. These properties can best be understood for bone or any other material by examining the material under loading. When a load in a known direction is placed on a structure, the deformation of that structure can be measured and plotted on a load-deformation curve, and the strength and stiffness of the structure can be determined.

A hypothetical load-deformation curve for a somewhat pliable material is shown in Figure 1–1. When a load is applied within the elastic region of the curve and is then released, the structure returns to its original shape; that is, no permanent deformation occurs. If loading is continued, the outermost fibers of the material begin to yield at some point. If loading continues past this yield point and into the nonelastic (plastic) region of the curve, permanent deformation results. If loading in the nonelastic region continues, an ultimate failure point for the structure is reached.

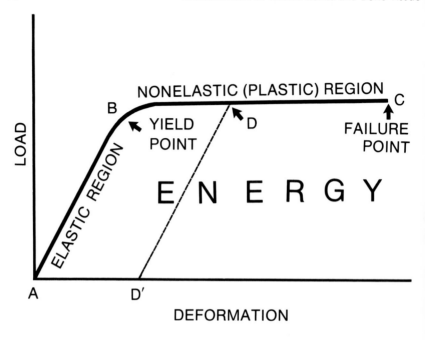

Fig. 1–1.   Load-deformation curve for a somewhat pliable material. If a load is applied in the elastic region (A–B) and is then released, no permanent deformation will occur. If loading is continued past the yield point and into the nonelastic (plastic) region (B–C) and the load is then released, permanent deformation will result. The amount of permanent deformation that will occur if the structure is loaded to point D and then unloaded is represented by the distance between A and D′. If loading continues in the nonelastic region, an ultimate failure point is reached.

The load-deformation curve shows three parameters for determining the strength of a structure: (1) the load the structure can sustain before failure, (2) the deformation it can sustain before failure, and (3) the energy it can store before failure. On the curve, the strength in terms of load and deformation is indicated by the ultimate failure point; the strength in terms of energy storage is indicated by the size of the area under the entire curve. In addition, the stiffness of the structure is indicated by the slope of the curve in the elastic region.

The load-deformation curve is useful for determining the strength and stiffness of whole structures of various sizes, shapes, and material composition. To examine the mechanical behavior of the material that composes a structure, and to compare the mechanical behavior of different materials, one must standardize the test specimens and the testing conditions. When samples of standardized size and shape are tested, the load per unit area and the amount of deformation in terms of length can be determined. The curve that is generated is called a stress-strain curve.

Stress is the load per unit area that develops on a plane surface within a structure in response to externally applied loads. It is expressed in force per unit area. The three units most commonly used for measuring stress in bone samples are newtons per centimeter squared (N/cm²); newtons per meter squared, or pascals (N/m²; Pa); and meganewtons per meter squared, or megapascals (MN/m²; MPa).

Strain is the deformation that occurs at a point in a structure under loading. Two basic types of strain exist: normal strain, which is a change in length, and shear strain, which is a change in angle. Normal strain is the amount of deformation (lengthening or shortening) divided by the structure's

Fig. 1–2. Standardized bone specimen in a testing machine. The strain between the two gauge arms is measured with a strain gauge. The stress is calculated from the total load measured. (Courtesy of Dennis R. Carter, Ph.D.)

original length. It is a nondimensional parameter expressed as a percentage (for example, centimeter per centimeter). Shear strain is the amount of angular deformation that occurs in a structure, that is, the amount that the original angle of the structure changes. It is expressed in radians (one radian equals approximately 57.3 degrees).

Stress and strain values can be obtained for bone by placing a standard specimen of bone tissue in a testing jig and loading it to failure (Fig. 1–2). The deformation that results can be illustrated on a stress-strain curve (Fig. 1–3). The regions of the stress-strain curve are similar to those of the load-deformation curve. Loads in the elastic region do not cause permanent deformation. Once the yield point is exceeded, however, permanent deformation occurs. The stiffness of the material is represented by the slope of the curve in the elastic region. The strength in terms of energy storage is represented by the area under the entire curve.

Stress-strain curves for metal, glass, and bone illustrate the differences in mechanical behavior among these materials (Fig. 1–4). Variations in stiffness are reflected by differences in the slope of the curves in the elastic region. Metal has the steepest slope and is thus the stiffest material. A value for stiffness is obtained by dividing the stress at any point in the elastic portion of

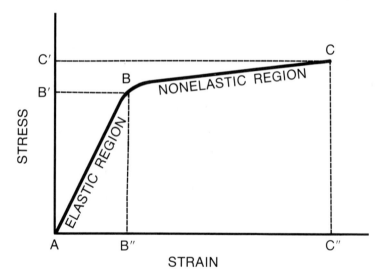

Fig. 1–3. Stress-strain curve for cortical bone tested in tension.
   Yield point (B)—point past which some permanent deformation of the bone occurred.
   Yield stress (B′)—load per unit area that the bone sample sustained before nonelastic deformation took place.
   Yield strain (B″)—amount of deformation that the sample sustained before nonelastic deformation occurred. The strain at any point in the elastic region of the curve is proportional to the stress at that point.
   Ultimate failure point (C)—the point past which failure of the sample occurred.
   Ultimate stress (C′)—load per unit area that the sample sustained before failure.
   Ultimate strain (C″)—amount of deformation that the sample sustained before failure.

the curve by the strain at that point. This value is called the modulus of elasticity (Young's modulus). Stiffer materials have higher moduli.

The elastic portion of the stress-strain curve for metal is a straight line, indicating linearly elastic behavior. Precise testing has shown that the elastic portion of the curve for bone is not straight, but curves slightly, indicating that bone is not linearly elastic in its behavior (Bonefield and Li, 1967). Some yielding may occur when the bone is loaded in the elastic region. The difference in nonelastic behavior for the two materials is due to differences in micromechanical events at yield. Yielding in bone (tested in tension) is caused by debonding of the osteons and microfracture. Yielding in metal (tested in tension) is caused by plastic flow and formation of plastic slip lines; slip lines are formed when the molecules of the lattice structure dislocate.

Materials are classified as brittle or ductile depending on the amount of deformation they undergo before failure. Glass, a typical brittle material, deforms little before failure, as indicated by the absence of a nonelastic (plastic) region on the stress-strain curve (see Fig. 1–4). Soft metal, a typical ductile material, deforms extensively before failure, as indicated by a long nonelastic (plastic) region on the curve (see Fig. 1–4). The fracture surfaces of the two types of material reflect this difference in amount of deformation (Fig. 1–5). A ductile material that is pieced together after fracture will not conform to its original shape, whereas a brittle material will.

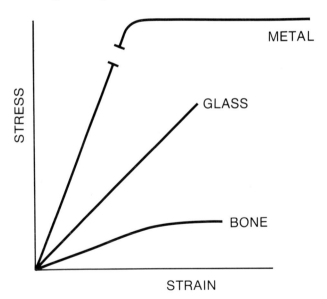

Fig. 1–4. Stress-strain curves for three materials. The stiffness of these materials is represented by the slope of the curve in the elastic region. The elastic portion of the curve for bone is slightly curved, indicating that bone is not linearly elastic in its behavior. Soft metal, a ductile material, has a long plastic region. Glass, a brittle material, has no plastic region.

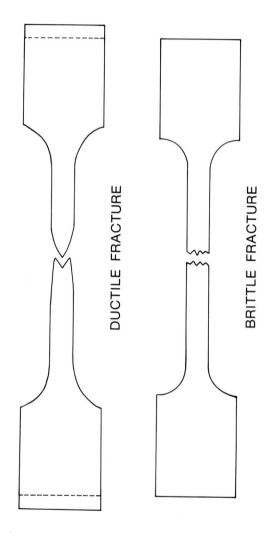

Fig. 1–5. Fracture surfaces of a ductile and a brittle material. The dotted lines on the ductile material indicate the original length of the structure, before deformation occurred. The brittle material deformed very little before fracture.

BONE TISSUE STRUCTURE

The skeleton is composed of cortical and cancellous bone. These two bone types can be considered as one material whose porosity varies over a wide range (Carter and Hayes, 1977b). Porosity is the proportion of the bone's volume occupied by nonmineralized (nonbone) tissue. It is expressed as a percentage. The difference in the porosity of the two bone types can be seen in the cross sections from human tibiae shown in Figure 1–6. The porosity varies from 5 to 30% in cortical bone and from 30 to over 90% in cancellous bone. The distinction between porous cortical bone and dense cancellous bone is somewhat arbitrary.

Cortical bone is stiffer than cancellous bone; it can withstand greater stress but less strain before failure. Cortical bone fractures in vitro when the strain exceeds 2% of the original length; cancellous bone does not fracture until the strain exceeds 7%. Because of its porous structure, cancellous bone has been shown to have a high energy storage capacity (Carter and Hayes, 1976).

Both cortical bone and cancellous bone are anisotropic. An anisotropic material exhibits different mechanical properties when loaded in different directions. Because the structure of bone is different in the transverse and the

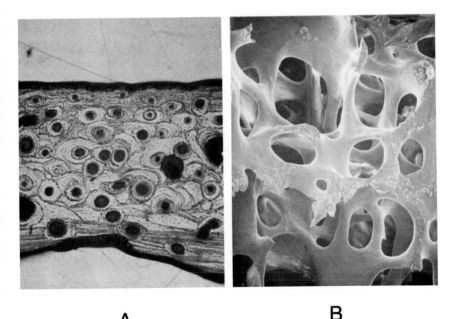

A   **B**

Fig. 1–6. **A.** Reflected-light photomicrograph of cortical bone from a human tibia (40×). (Courtesy of Dennis R. Carter, Ph.D.)
   B. Scanning electron photomicrograph of cancellous bone from a human tibia (30×). (Courtesy of Dennis R. Carter, Ph.D.)

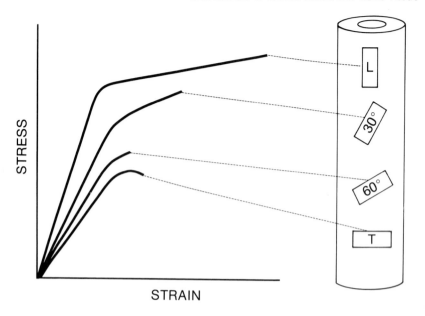

Fig. 1–7. Anisotropic behavior of cortical bone specimens from a human femoral shaft tested
in tension in four directions: longitudinal (L), tilted 30 degrees with respect to the neutral
axis of the bone, tilted 60 degrees, and transverse (T). (Data from Frankel and Burstein,
1970.)

longitudinal directions, bone strength varies according to the direction in
which the load is imposed. The strength and stiffness would be expected to
be the greatest in the direction in which loads are most commonly imposed
on the bone. Figure 1–7 shows the variations in strength and stiffness for
cortical bone samples from a human femoral shaft, tested in tension in four
directions (Frankel and Burstein, 1970). The values for both parameters are
highest for the samples loaded in the longitudinal direction.

BONE BEHAVIOR UNDER VARIOUS LOADING MODES

Forces and moments can be applied to a structure in various directions,
producing tension, compression, bending, shear, torsion, and combined
loading (Fig. 1–8). The following descriptions of these loading modes apply
to structures in equilibrium.

Tension

In tensile loading, equal and opposite loads are applied outward from the
surface of the structure, and tensile stress and strain result inside the
structure. Tensile stress can be thought of as many small forces directed

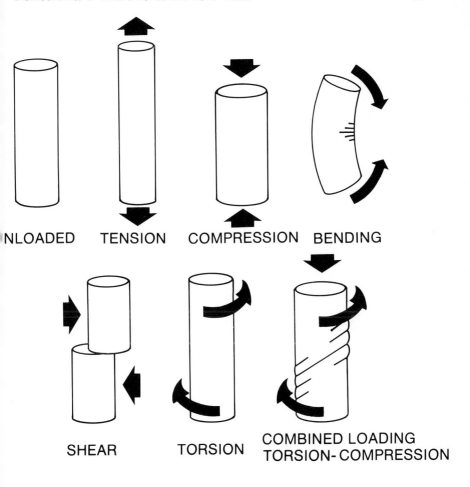

Fig. 1–8. Schematic representation of various loading modes.

away from the surface of the structure. Maximal tensile stress occurs on a plane perpendicular to the applied load (Fig. 1–9). Under tensile loading, the structure lengthens and narrows. The failure mechanism of bone tissue loaded in tension is mainly debonding at the cement lines and pulling out of the osteons (Fig. 1–10).

Clinically, fractures produced by tensile loading are usually seen in cancellous bone. Examples are fractures of the base of the fifth metatarsal adjacent to the attachment of the peroneus brevis tendon and fractures of the calcaneus adjacent to the attachment of the Achilles tendon. Figure 1–11 shows a tensile fracture through the calcaneus; strong contraction of the triceps surae muscle produced abnormally high tensile loads on this bone.

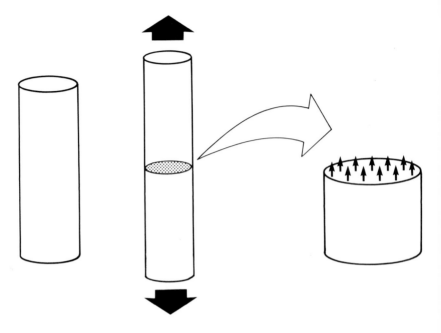

Fig. 1–9. Tensile loading.

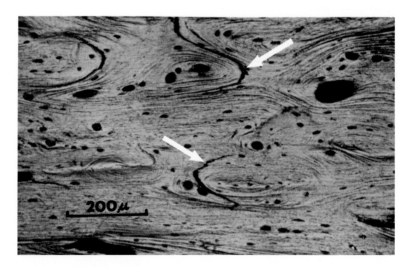

Fig. 1–10. Reflected-light photomicrograph of a human cortical bone specimen tested in tension (30×). Arrows indicate debonding at the cement lines and pulling out of the osteons. (Courtesy of Dennis R. Carter, Ph.D.)

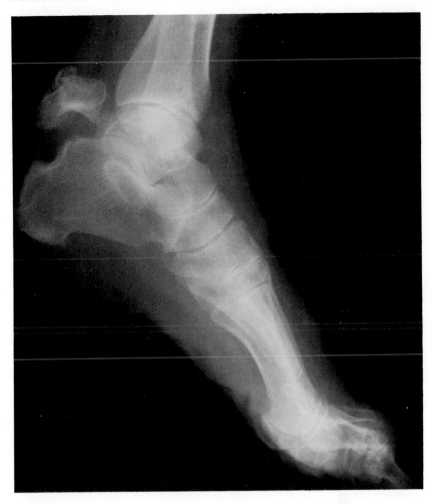

Fig. 1–11. Tensile fracture through the calcaneus produced by strong contraction of the triceps surae muscle. (Courtesy of Robert A. Winquist, M.D.)

Compression

In compressive loading, equal and opposite loads are applied toward the surface of the structure, and compressive stress and strain result inside the structure. Compressive stress can be thought of as many small forces directed into the surface of the structure. Maximal compressive stress occurs on a plane perpendicular to the applied load (Fig. 1–12). Under compressive loading, the structure shortens and widens. The failure mechanism for bone tissue loaded in compression is mainly oblique cracking of the osteons (Fig. 1–13).

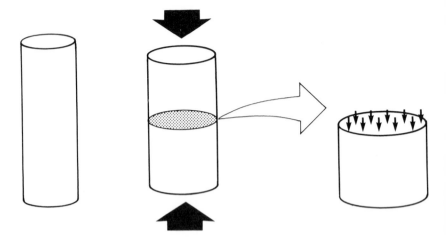

Fig. 1–12. Compressive loading.

Fig. 1–13. Scanning electron photomicrograph of a human cortical bone specimen tested in compression (30×). Arrows indicate oblique cracking of the osteons. (Courtesy of Dennis R. Carter, Ph.D.)

Fractures produced by compressive loading are usually found in the vertebrae. Figure 1–14 shows the shortening and widening that took place in a human vertebra subjected to a high compressive load. Compressive loading to failure in a joint can be produced by an abnormally strong contraction of the surrounding muscles. An example of this effect is presented in Figure 1–15: bilateral subcapital fractures of the femoral neck occurred during electric shock therapy; contraction of the muscles around the hip joint produced compression of the femoral head against the acetabulum.

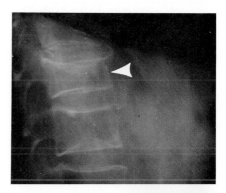

Fig. 1–14. Compressive fracture of a human vertebra at the L1 level. The vertebra has shortened and widened.

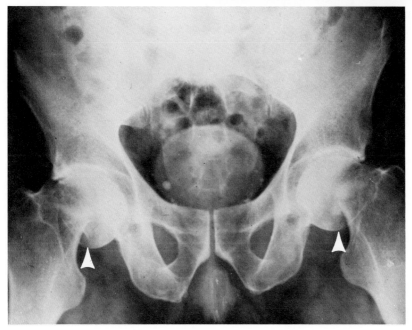

Fig. 1–15. Bilateral subcapital compression fractures of the femoral necks in a patient who underwent electric shock therapy.

Shear

In shear loading, a load is applied parallel to the surface of the structure, and shear stress and strain result inside the structure. Shear stress can be thought of as many small forces acting on the surface of the structure on a plane parallel to the applied load (Fig. 1–16).

When subjected to shear load, the structure deforms internally in an angular manner; right angles become obtuse or acute (Fig. 1–17). Figure 1–18 illustrates angular deformation in a structure subjected to tensile and compressive loading.

Shear fractures are usually seen in cancellous bone. Examples are fractures of the femoral condyles and the tibial plateau. A shear fracture of the tibial plateau is shown in Figure 1–19.

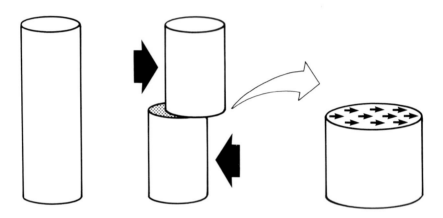

Fig. 1–16. Shear loading.

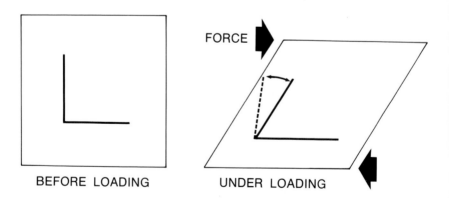

BEFORE LOADING   UNDER LOADING

Fig. 1–17. When a structure is loaded in shear, lines originally at right angles change their orientation, and the angle becomes obtuse or acute. This change in angle indicates shear strain. (Adapted from Frankel and Burstein, 1970.)

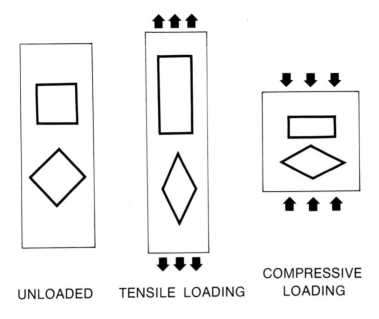

Fig. 1–18. The presence of shear strain in a structure loaded in tension and in compression is indicated by angular deformation. (Adapted from Frankel and Burstein, 1970.)

The ultimate stress for human adult cortical bone differs under compressive, tensile, and shear loading (Fig. 1–20). Cortical bone can withstand greater stress in compression than in tension, and greater stress in tension than in shear (Reilly and Burstein, 1975). In shear loading the value for the stiffness of the material is known as the shear modulus rather than the modulus of elasticity.

Bending

Bending occurs when a load is applied to a structure in a manner that causes it to bend about an axis. The structure under bending is subjected to a combination of tension and compression. When a bone is loaded in bending, tensile stresses and strains act on one side of the neutral axis, and compressive stresses and strains act on the other side (Fig. 1–21); there are no stresses and strains at the neutral axis. The magnitude of the stresses is proportional to their distance from the neutral axis of the bone. The farther the stresses are from the neutral axis, the higher their magnitude. Because bone is asymmetrical, the tensile and compressive stresses may not be equal.

Two types of bending exist: bending produced by three forces (three-point bending) and that produced by four forces (four-point bending) (Fig. 1–22). Fractures produced by both types of bending are commonly observed.

Three-point bending takes place when three forces acting on a structure

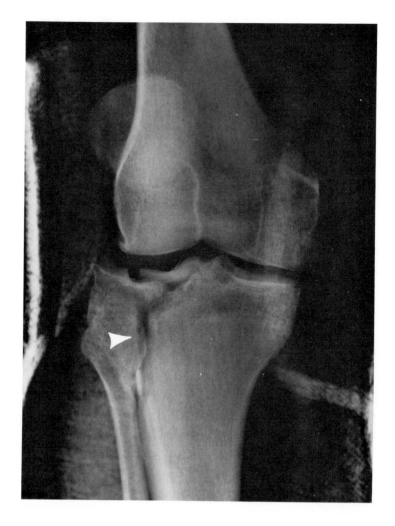

Fig. 1–19. Shear fracture of the lateral tibial plateau. (Courtesy of Robert A. Winquist, M.D.)

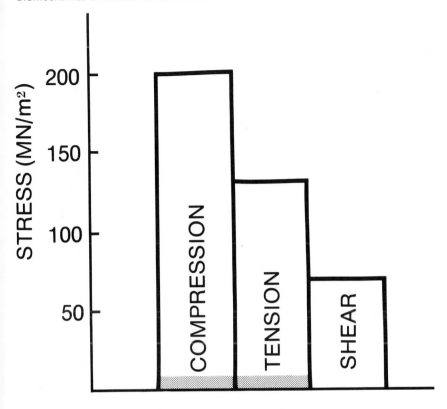

Fig. 1–20. Ultimate stress for human adult cortical bone specimens tested in compression, tension, and shear. (Average of data from Reilly and Burstein, 1975.) Shaded area indicates ultimate stress for human adult cancellous bone with an apparent density of 0.35 tested in tension and compression (Carter, 1979).

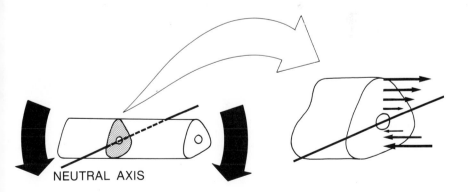

Fig. 1–21. Cross section of a bone subjected to bending, showing distribution of stresses around the neutral axis. Tensile stresses act on the superior side, and compressive stresses act on the inferior side. The magnitude of the stresses is highest at the periphery of the bone and lowest near the neutral axis. The tensile and compressive stresses are not equal because the bone is asymmetrical.

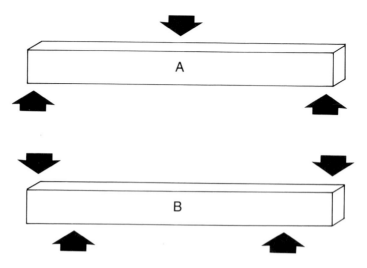

Fig. 1–22. Two types of bending: **A.** Three-point bending.
B. Four-point bending.

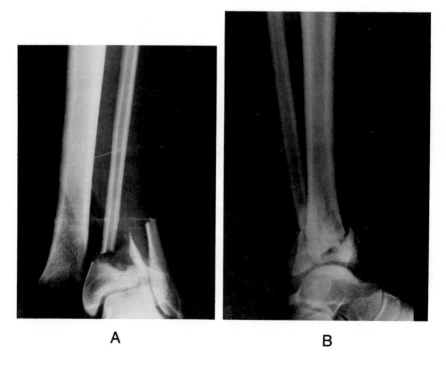

A   B

Fig. 1–23. Anteroposterior **(A)** and lateral **(B)** roentgenograms of a ''boot top'' fracture
produced by three-point bending. (Courtesy of Robert A. Winquist, M.D.)

produce two equal moments (see Fig. 1–22A). The structure will break at the point of application of the middle force.

A typical three-point bending fracture is the "boot top" fracture sustained during skiing. In the "boot top" fracture shown in Figure 1–23, one bending moment was applied to the upper part of the tibia as the skier fell forward over the top of the ski boot. An equal moment was produced by the fixed foot and ski. As the upper part of the tibia was bent forward, tensile stresses and strains acted on the posterior side of the bone, and compressive stresses and strains acted on the anterior side. In adult bone, failure begins on the side subjected to tension, since adult bone is weaker in tension than in compression. Immature bone may fail first in compression, and a buckle fracture may result on the compressive side.

Four-point bending takes place when two force couples acting on a structure produce two equal moments. A force couple is formed when two

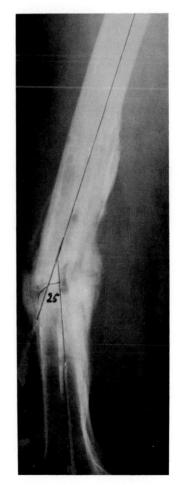

Fig. 1–24. Lateral roentgenogram of a femoral fracture produced by four-point bending. Manipulation of the knee during rehabilitation caused the femur to refracture at its weakest point—the original fracture site. (Courtesy of Kaj Lundborg, M.D.)

parallel forces of equal magnitude but opposite direction are applied to a structure (see Fig. 1–22B). Because the magnitude of the bending moment is the same throughout the area between the two force couples, the structure will break at its weakest point.

An example of a four-point bending fracture is shown in Figure 1–24. A stiff knee joint was manipulated incorrectly during rehabilitation of a femoral fracture. As the knee was manipulated, the posterior knee capsule and tibia formed one force couple, and the femoral head and hip joint capsule formed the other. As a bending moment was applied to the femur, it failed at its weakest point—the original fracture site.

Torsion

Torsion occurs when a load is applied to a structure in a manner that causes it to twist about an axis. When a structure is loaded in torsion, shear stresses are distributed over the entire structure. As in bending, the magnitude of these stresses is proportional to their distance from the neutral axis (Fig. 1–25). The farther the stresses are from the neutral axis, the higher their magnitude.

Under torsional loading, maximal shear stresses act on planes parallel and perpendicular to the neutral axis of the structure. In addition, maximal tensile and compressive stresses act on a plane diagonal to the neutral axis of the structure. Figure 1–26 illustrates these planes in a small segment of bone loaded in torsion.

The fracture pattern for bone loaded in torsion suggests that the bone fails first in shear, with the formation of a crack parallel to the neutral axis of the bone. Usually, a crack then forms along the plane of maximal tensile stress. Such a pattern can be seen in the experimentally produced torsional fracture of a dog femur shown in Figure 1–27.

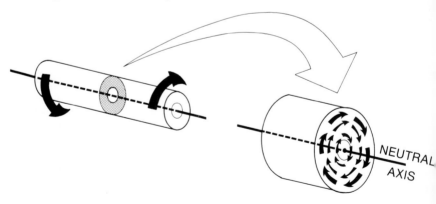

Fig. 1–25. Cross section of a cylinder loaded in torsion, showing the distribution of shear stresses around the neutral axis. The magnitude of the stresses is highest at the periphery of the cylinder and lowest near the neutral axis.

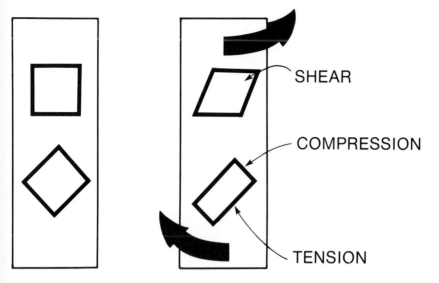

Fig. 1–26. In a small segment of bone loaded in torsion, maximal shear stresses act on planes parallel and perpendicular to the neutral axis. Maximal tensile and compressive stresses act on planes diagonal to the neutral axis.

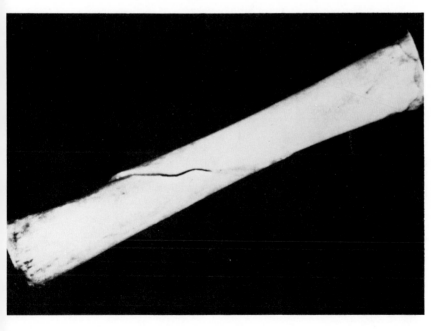

Fig. 1–27. Experimentally produced torsional fracture of a dog femur. The short crack parallel to the neutral axis represents shear failure; the fracture line at 30 degrees to the neutral axis represents the plane of maximal tensile stress.

Combined Loading

Although each loading mode has been discussed separately, living bone is seldom loaded in only one mode. Loading of bone in vivo is complex for two main reasons: the geometric structure of the bone is irregular, and bone is constantly subjected to multiple indeterminate loads.

Recent in vivo measurement of the strains on the anteromedial surface of a human adult tibia during walking and jogging demonstrated the complexity of loading during these common physiologic activities (Lanyon et al.,1975). Carter (1978) calculated stress values from these strain measurements. During normal walking the stresses were compressive during heel strike, tensile during the stance phase, and then compressive during push off (Fig. 1–28A). Relatively high shear stress appeared in the later portion of the gait cycle, denoting significant torsional loading. This torsional loading indicated external rotation of the tibia during stance and push off.

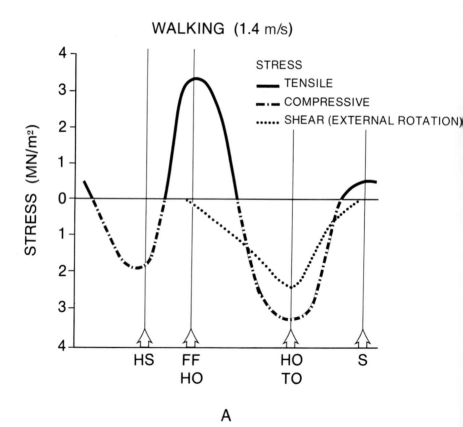

A

During jogging the stress pattern was quite different (Fig. 1–28B). The stress at toe strike was primarily compressive. This was followed by high tensile stress during push off. The shear stress was small throughout the stride, denoting minimal torsional loading. This torsional loading indicated both external and internal rotation of the tibia in an alternating pattern.

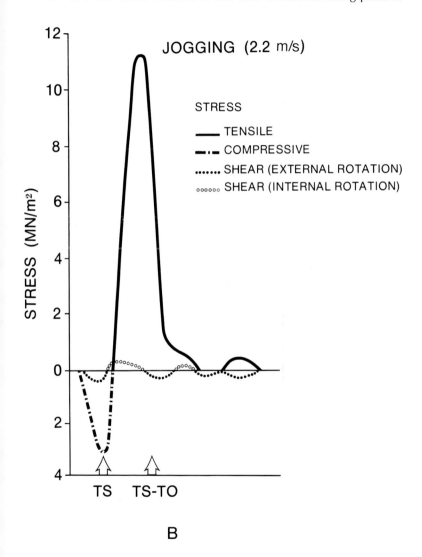

B

Fig. 1–28. **A.** Calculated stresses on the anterior lateral cortex of a human adult tibia during walking. HS—heel strike; FF —full foot; HO—heel off; TO —toe off; S—swing. (After Lanyon et al., 1975.) (Courtesy of Dennis R. Carter, Ph.D.)
  B. Calculated stresses on the anterior lateral cortex of an adult human tibia during jogging. TS—toe strike; TO—toe off. (After Lanyon et al., 1975.) (Courtesy of Dennis R. Carter, Ph.D.)

Clinical examination of fracture patterns indicates that few fractures are produced by one loading mode or even by two modes; most fractures are produced by a combination of several loading modes.

INFLUENCE OF MUSCLE ACTIVITY ON STRESS DISTRIBUTION IN BONE

When bone is loaded in vivo, contraction of the muscles attached to the bone alters the stress distribution in the bone. This muscle contraction decreases or eliminates tensile stress on the bone by producing compressive stress that either partially or totally neutralizes it.

The effect of muscle contraction can be illustrated in a tibia subjected to three-point bending. Figure 1–29A represents the leg of a skier who is falling forward, subjecting his tibia to a bending moment. High tensile stress is produced on the posterior aspect of the tibia, and high compressive stress acts on the anterior aspect. Contraction of the triceps surae muscles

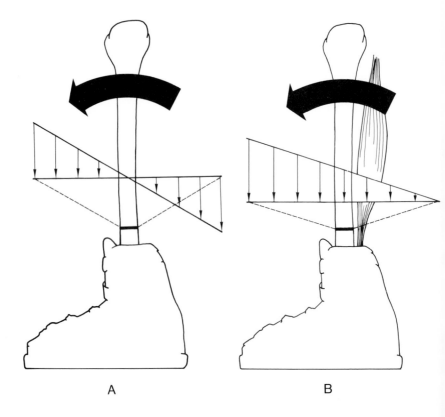

A   B

Fig. 1–29. A. Distribution of compressive and tensile stresses in a tibia subjected to three-point bending.
  B. Contraction of the triceps surae muscle produces high compressive stress on the posterior aspect, neutralizing the high tensile stress.

produces a high compressive stress on the posterior aspect (Fig. 1–29B), neutralizing the high tensile stress and thereby protecting the tibia from failure in tension. This muscle contraction may result in a higher compressive stress on the anterior surface of the tibia. Adult bone can usually withstand this stress, but immature bone, which is weaker, may fail in compression.

Muscle contraction produces a similar effect in the hip joint (Fig. 1–30). During locomotion, bending moments are applied to the femoral neck, and tensile stress is produced on the superior cortex. Contraction of the gluteus

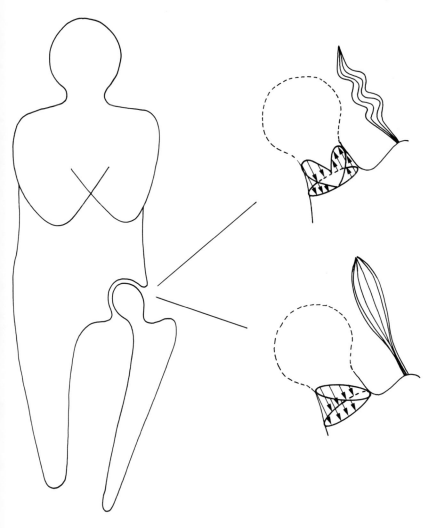

Fig. 1–30. Distribution of stresses in a femoral neck subjected to bending. When the gluteus medius muscle is relaxed (top), tensile stress acts on the superior cortex and compressive stress acts on the inferior cortex. Contraction of this muscle (bottom) neutralizes the tensile stress.

medius muscle produces compressive stress that neutralizes this tensile stress, with the net result that neither compressive nor tensile stress acts on the superior cortex. Thus, the muscle contraction allows the femoral neck to sustain higher loads than would otherwise be possible.

ENERGY STORAGE IN BONE

During normal activity a small proportion of the total energy storage capacity of bone is utilized. To illustrate how little of this capacity is utilized during a normal physiologic activity (jogging), the tensile strain values from the Lanyon experiment (Lanyon et al.,1975) have been plotted on a stress-strain curve for human cortical bone samples tested to failure in tension (Fig. 1–31).

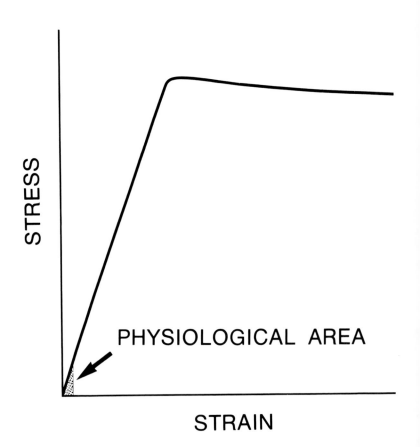

Fig. 1–31.   Tensile strain values from a human adult tibia during jogging (Lanyon et al., 1975) have been plotted on a stress-strain curve for bone samples tested to failure in tension. A small proportion of the total energy storage capacity of the bone is utilized during this normal physiologic activity.

The energy storage capacity of bone varies with the speed at which it is loaded. The higher the speed of loading, the more energy the bone stores to failure. The load-deformation curves in Figure 1–32 show the difference in the energy storage capacity of paired dog tibiae tested in vitro at a high and a low loading speed (Sammarco et al., 1971). The amount of energy stored to failure approximately doubled when the fracture time increased from 0.01 second to 200 seconds. The load to failure almost doubled with the increase in speed, but the deformation to failure did not change significantly. The bone was stiffer at the higher speed.

The speed of loading is clinically significant because it influences both the fracture pattern and the amount of soft tissue damage at fracture. When a bone fractures, the stored energy is released. At a low loading speed, the energy can dissipate through the formation of a single crack; the bone and soft tissues remain relatively intact, and little or no displacement takes place. At a high loading speed, however, the greater energy stored cannot dissipate rapidly enough through a single crack, and comminution and extensive soft tissue damage occur. Figure 1–33 shows a tibia tested in vitro in torsion at a high loading speed; numerous bone fragments were produced, and displacement of the bone was pronounced.

Fractures fall into three general categories based on the amount of energy released at fracture: low-energy, high-energy, and very high energy. The low-energy fracture is exemplified by the simple torsional ski fracture; the high-energy fracture is often sustained during automobile accidents; and the very high energy fracture is produced by gunshot of a very high muzzle velocity.

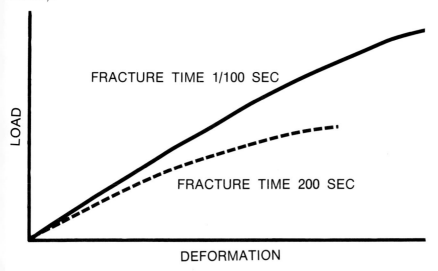

Fig. 1–32. Energy storage in paired dog tibiae tested at a high and a low loading speed. The load to failure and the energy stored to failure almost doubled at the higher speed. (Adapted from Sammarco et al., 1971.)

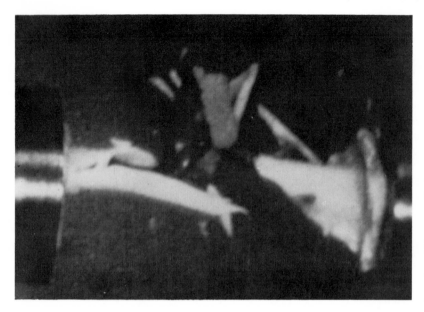

Fig. 1–33.   Tibia experimentally tested to failure in torsion at a high loading speed. Numerous fragments were produced, and displacement was pronounced.

FATIGUE OF BONE UNDER REPETITIVE LOADING

Fractures can be produced by either a single load or the repeated application of a load. A fracture will result from the application of a single load if that load exceeds the ultimate strength of the bone. The repeated application of a lower load may also produce a fracture; such a fracture is called a fatigue fracture. A fatigue fracture is typically produced by either low repetition of high loads or high repetition of relatively normal loads.

The interplay of load and repetition for all materials can be plotted on a fatigue curve (Fig. 1–34). For some materials (some metals, for example), the fatigue curve is asymptotic, indicating that if the load is kept below a certain level, the material will remain intact, no matter how large the number of repetitions. For bone tested in vitro, the curve may not be asymptotic. Fatigue microfractures may be created in bone subjected to low repetitive loads (Carter and Hayes, 1977a).

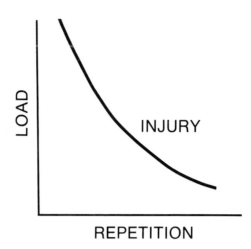

Fig. 1–34. The interplay of load and repetition is represented on a fatigue curve.

Testing of bone in vitro also reveals that bone fatigues rapidly when load or deformation approaches the yield strength of the bone (Carter and Hayes, 1977a); that is, the number of repetitions needed to produce a fracture greatly diminishes.

In repetitive loading of living bone, not only do the amount of load and the number of repetitions affect the fatigue process, but also the frequency of loading. Since living bone is self-repairing, a fatigue fracture only results when the remodeling process is out-paced by the fatigue process, that is, when the frequency of loading precludes the remodeling necessary to prevent failure.

Fatigue fractures are usually sustained during continuous strenuous physical activity. Such activity causes the muscles to fatigue. When muscles fatigue, their ability to contract is reduced; as a result they are less able to store energy and thus to neutralize the stresses imposed on the bone. The resulting alteration of the stress distribution in the bone causes abnormally high loads to be imposed on the bone, and a fatigue fracture may be produced. Failure may occur on the tensile side, the compressive side, or both sides of the bone. Failure on the tensile side results in a transverse crack, and the bone may proceed rapidly to complete fracture. Fatigue fractures on the compressive side appear to be produced more slowly; the remodeling is less easily out-paced by the fatigue process, and the bone may not proceed to complete fracture.

This theory of muscle fatigue as a cause of fatigue fracture in the lower extremities is outlined in the following schema:

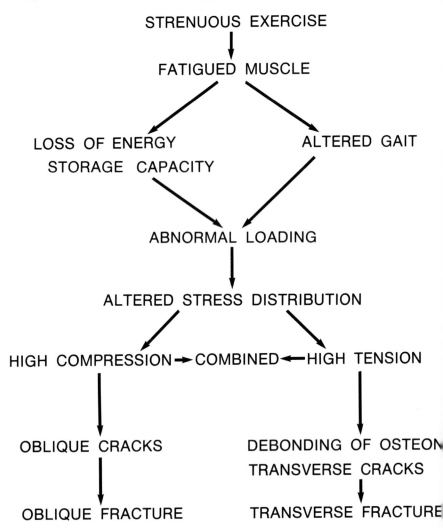

INFLUENCE OF BONE SIZE AND SHAPE ON STRENGTH AND STIFFNESS

The size and shape of a bone play a major role in determining its strength and stiffness.

In tension and compression, the cross-sectional area (size) of a bone affects its strength and stiffness. The larger the area is, the greater the strength and stiffness. Both the load to failure and the stiffness are proportional to the cross-sectional area.

In bending, both the cross-sectional area (size) and the distribution of bone tissue around a neutral axis (shape) affect strength and stiffness. The quantity that takes into account the area and the distribution of material around an axis in bending is the area moment of inertia. The larger the area moment of inertia is, the stronger and stiffer the bone.

Figure 1–35 shows the influence of the area moment of inertia on load to failure and stiffness of three rectangular structures: these beams have the same area but different shapes. In bending, Beam III is the stiffest of the three and can withstand the greatest load, because the greatest amount of material is distributed away from the neutral axis. For rectangular cross sections, the formula for the area moment of inertia is the width (B) multiplied by the cube of the height (H^3) divided by 12: $\frac{B\ H^3}{12}$. Because of its large area moment of inertia, Beam III can withstand four times more load in bending than Beam I.

A third factor, the length of the bone, influences the strength and stiffness in bending. The longer the bone is, the greater the magnitude of the bending moment caused by the application of a force. In a rectangular structure, the magnitude of the stresses produced at the point of application of the bending moment is proportional to the length of the structure. Figure 1–36 depicts the forces acting on two beams with the same width and height but different lengths: Beam B is twice as long as Beam A. The bending moment for the longer beam is twice that for the shorter beam; consequently, the stress magnitude throughout the beam is twice as high.

Because of their length, the long bones of the skeleton are subjected to high bending moments, and thus high tensile and compressive stresses. Their tubular shape gives them the ability to resist bending moments in all directions. The bones have a large area moment of inertia, because much of the bone tissue is distributed away from the neutral axis.

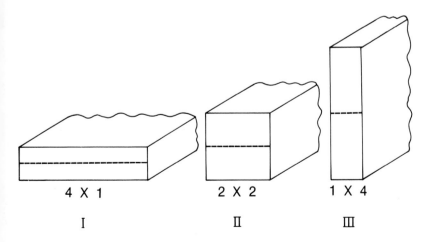

4 X 1   2 X 2   1 X 4

I   II   III

Fig. 1–35. Three beams of equal area but different shapes. For rectangular cross sections, the area moment of inertia is calculated by the formula $\frac{B\ H^3}{12}$, where B is the width and H, the height. The area moment of inertia for Beam I is 4/12; for Beam II, 16/12; and for Beam III, 64/12. (Adapted from Frankel and Burstein, 1970.)

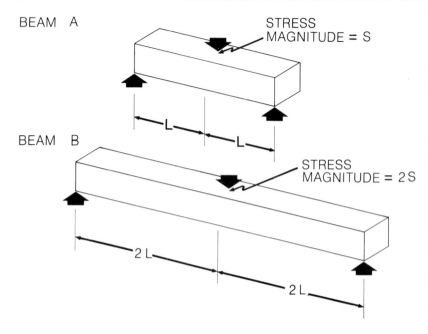

Fig. 1–36. Beam B is twice as long as Beam A and has twice the bending moment. Therefore, the stress magnitude throughout Beam B is also twice as high. (Adapted from Frankel and Burstein, 1970.)

In torsion, two factors affect bone strength and stiffness: the cross-sectional area (size) and the distribution of bone tissue around a neutral axis (shape). The quantity that takes into account these two factors in torsional loading is the polar moment of inertia. The larger the polar moment of inertia is, the stronger and stiffer the bone.

Figure 1–37 shows distal and proximal cross sections of a tibia subjected to torsional loading. Although it has a slightly smaller bony area, the proximal section has a much higher polar moment of inertia, because much of the bone tissue is distributed away from the neutral axis. The distal section, while having a larger bony area, is subjected to a much higher shear stress because much of the bone tissue is distributed near the neutral axis. The magnitude of the shear stress in the distal section is approximately double that in the proximal section. Torsional fractures of the tibia commonly occur distally, as would be expected.

A number of processes affect the size, shape, and tissue structure of bone, and thus its mechanical behavior. The most common of these are fracture healing, surgical alterations of bone, bone remodeling, and aging.

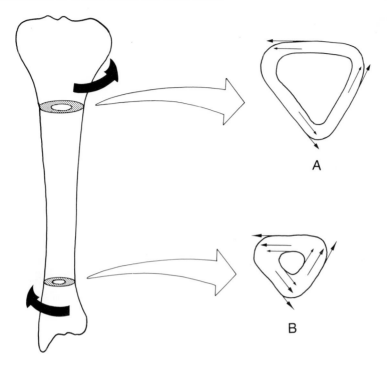

Fig. 1–37. Distribution of shear stresses in two cross sections of a tibia subjected to torsional loading. The proximal section (A) has a higher moment of inertia than the distal section (B), because more bony material is distributed away from the neutral axis. (Adapted from Frankel and Burstein, 1970.)

Fracture Healing

When bone begins to heal after fracture, a cuff of callus forms around the fracture site to stabilize this area (Fig. 1–38A). The callus significantly increases the area and polar moments of inertia, thereby increasing the strength and stiffness of the bone, particularly in bending and torsion, during the healing period. As the fracture heals and the bone gradually regains its normal strength, the cuff is progressively resorbed, and the bone returns as closely as possible to its normal size and shape (Fig. 1–38B).

Surgical Alterations of Bone

Certain surgical procedures produce defects that greatly weaken the bone, particularly in torsion. These defects fall into two categories: those whose length is less than the diameter of the bone, and those whose length exceeds the bone diameter.

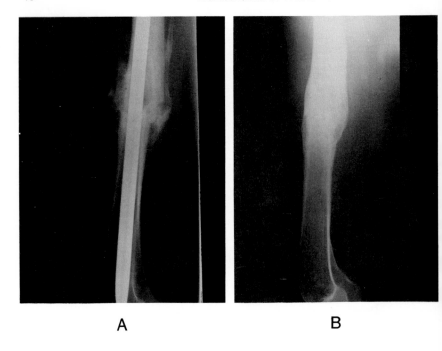

A   B

Fig. 1–38. **A.** Early callus formation in a femoral fracture fixed with an intramedullary nail. **B.** Nine months after fracture most of the callus cuff has been resorbed. (Courtesy of Robert A. Winquist, M.D.)

The smaller defect, a stress raiser, is produced surgically when a small piece of bone is removed or a screw is inserted in the bone. A stress raiser reduces the strength of the bone by preventing the stresses imposed during loading from being distributed evenly throughout the bone. Instead, these stresses become concentrated around the defect. This defect is analogous to a rock in a stream, which diverts the water, producing a high water flow around it (Fig. 1–39). The weakening effect of a stress raiser is particularly marked under torsional loading; the total decrease in bone strength in this loading mode can be as much as 60%.

Burstein et al. (1972) showed the effect of screws and screw holes on the energy storage capacity of rabbit bones. The immediate effect of inserting a screw and drilling a hole in a rabbit femur was a 70% decrease in energy storage capacity. After 8 weeks, the stress raiser effect produced by the hole and the screw had disappeared completely due to remodeling of the bone. However, when the screw was removed, this effect again immediately developed because of microdamage produced by screw removal (Fig. 1–40).

The larger defect, an open section defect, is a discontinuity in the bone caused by surgical removal of a piece of bone longer than the bone's diameter.

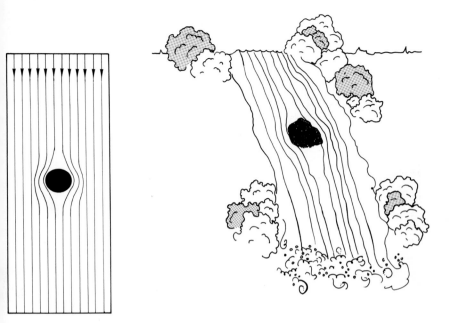

Fig. 1–39. Stress concentration around a defect; such a defect is analogous to a rock in a stream.

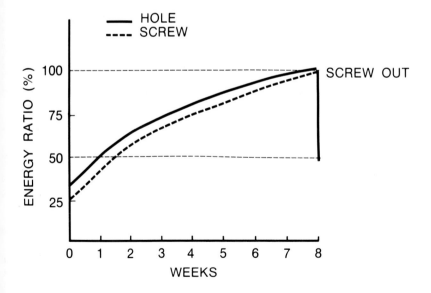

Fig. 1–40. Effect of screws and screw holes on the energy storage capacity of rabbit femora. The energy storage for experimental animals is expressed as a percentage of the total energy storage capacity for control animals. (Adapted from Burstein et al., 1972.)

In normal bone subjected to torsional loading, the shear stress is distributed throughout the bone and acts to resist the torque. This stress pattern is illustrated in the cross section of a long bone shown in Figure 1–41A. A cross section with a continuous outer surface is called a closed section.

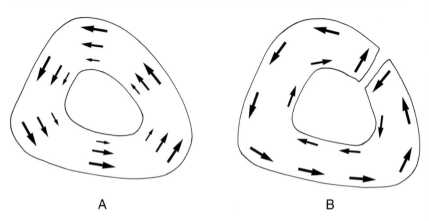

Fig. 1–41.   Stress pattern in an open and closed section under torsional loading.
A. In the closed section, the shear stress resists the applied torque.
B. In the open section, only the shear stress at the periphery of the bone resists the applied torque. (Adapted from Frankel and Burstein, 1970.)

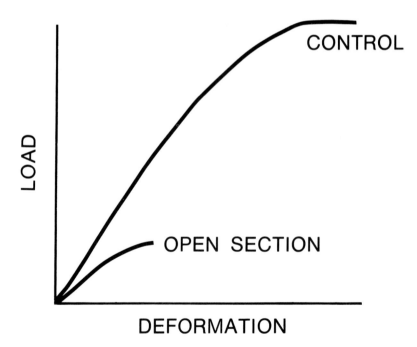

Fig. 1–42.   Load-deformation curves for human adult tibiae tested in vitro under torsional loading. The control curve represents a tibia with no defect; the open section curve represents a tibia with an open section defect. (Adapted from Frankel and Burstein, 1970.)

Surgically removing a large piece of bone (for example, by cutting a slot) creates an open section, in which the outer surface of the bone cross section is no longer continuous. In bone with an open section defect, only the shear stress at the periphery of the bone resists the applied torque. As the shear stress encounters the discontinuity, it is forced to change direction (Fig. 1–41B). Throughout the interior of the bone, the stress runs parallel to the applied torque, and the amount of bone tissue resisting the load is greatly decreased.

In torsional tests of human adult tibiae in vitro, an open section defect reduced the load to failure and energy storage to failure by up to 90%. The deformation to failure was diminished by about 70% (Frankel and Burstein, 1970) (Fig. 1–42).

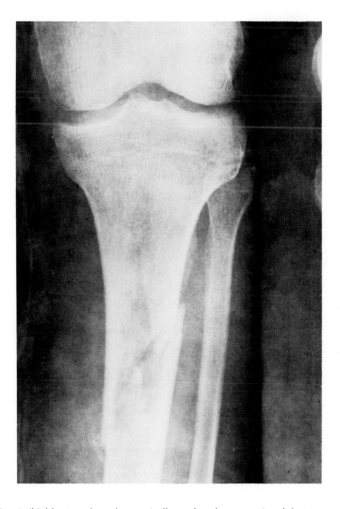

Fig. 1–43. A tibial fracture through a surgically produced open section defect occurred when the patient tripped a few weeks after operation.

Clinically, bone can be greatly weakened after surgical removal of a piece of bone. Figure 1–43 presents a roentgenogram of a tibia from which a graft was removed for use in an arthrodesis of the hip. The open section defect weakened the tibia, particularly in torsion. A few weeks after operation, the patient tripped, and a fracture occurred through the defect.

Bone Remodeling

Bone remodeling is the ability of bone to adapt, by changing its size, shape, and structure, to the mechanical demands placed upon it. This adaptation takes place according to Wolff's law, which states that bone is laid down where needed and resorbed where not needed.

If, because of decreased mobilization or immobilization, bone is not subjected to the usual mechanical stresses, resorption of periosteal and subperiosteal bone takes place (Jenkins and Cochran, 1969), and strength

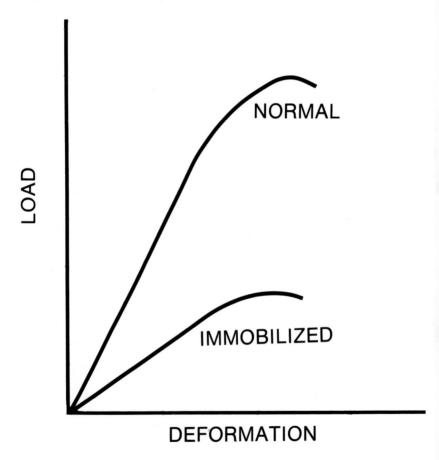

Fig. 1–44. Load-deformation curves for vertebral segments (L5-L7) from normal and immobilized Rhesus monkeys. (Adapted from Kazarian and Von Gierke, 1969.)

and stiffness decrease. This decrease in strength and stiffness of bone was shown by Kazarian and Von Gierke (1969), who immobilized Rhesus monkeys in full body casts for a 60-day period. Subsequent in vitro compressive testing of the vertebrae from the immobilized monkeys and from controls showed up to a threefold decrease in load to failure and energy storage capacity in the immobilized vertebrae; stiffness was also significantly decreased (Fig. 1–44).

Remodeling of a bone following fracture healing is influenced by the presence of an implant which remains firmly attached to the bone, for example, a plate fixed to the bone with screws. In this situation there will be load sharing between the plate and the bone. The proportion of the loads borne by each structure will depend upon the geometry and material properties of the structure. Bone will remodel, that is, add or lose cortical and/or cancellous bone, depending upon the level of stress it sustains. A large plate, carrying high loads, will in effect unload the bone. In response to this diminished load the bone will atrophy. In the process of transferring loads from the plate to the bone, the bone-screw interface will carry large loads, and hypertrophy of the bone will occur at the interface.

An example of bone resorption under a plate is illustrated in Figure 1–45. A compression plate, composed of a material approximately ten times stiffer

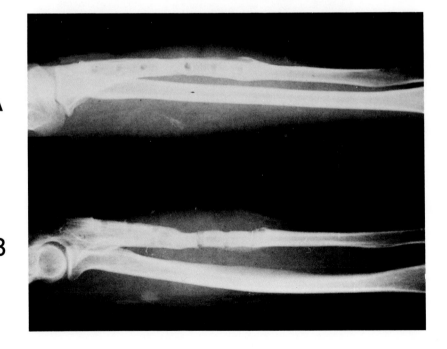

Fig. 1–45. Anteroposterior **(A)** and lateral **(B)** roentgenograms of an ulna after plate removal show a decreased bone diameter due to resorption of the bone under the plate. Cancellization of the cortex and the presence of screw holes also contribute to weakening of the bone. (Courtesy of Marc Martens, M.D.)

than the bone, was applied to a fractured ulna and remained after the fracture had healed. The bone under the plate, carrying a lower load than normal, was partially resorbed, and the diameter of the diaphysis became markedly smaller. Such a diminution in the size of the bone diameter produces a great decrease in strength, particularly in bending and torsion, as it reduces the area and polar moments of inertia. A 20% decrease in bone diameter may reduce the strength in torsion by 60%.

An example of bone hypertrophy around screws is illustrated in Figure 1–46. A nail plate was applied to a femoral neck fracture. The bone hypertrophied around the screws in response to the increased load.

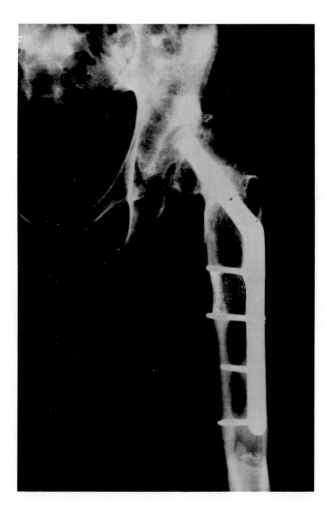

Fig. 1–46. Roentgenogram of a femoral neck fracture to which a nail plate was applied. The loads are transmitted from the plate to the bone via the screws. Bone has been laid down around the screws to bear this load.

If bone is repeatedly subjected to high mechanical stresses within the normal physiologic range, hypertrophy of periosteal and subperiosteal bone may result. Hypertrophy of normal adult bone as a response to strenuous exercise has been observed (Jones et al., 1977; Dalen and Olsson, 1974). An increase in bone density has also been noted (Nilsson and Westlin, 1971).

Degenerative Changes Due to Aging

As normal aging proceeds, the walls of the trabeculae in cancellous bone become progressively thinner, and many may even be resorbed (Fig. 1–47). The result is a marked reduction in the amount of cancellous bone and a decrease in the diameter and thickness of the cortex. This decrease in the total amount of bone tissue, as well as the slight decrease in the size of the bone, causes a reduction in strength and stiffness.

Stress-strain curves for specimens from human adult tibiae of two widely differing ages tested in tension are shown in Figure 1–48. The ultimate stress was approximately the same for the young and old bone. However, the old bone sample could withstand only half the strain of the young bone, indicating that the old bone was less ductile than the young and was able to store less energy to failure.

SUMMARY

1. The skeleton is composed of cortical and cancellous bone. These two bone types can be considered as one material whose porosity varies over a wide range.

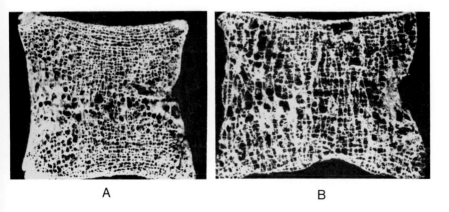

A　　　　　　　　　　　　B

Fig. 1–47. Vertebral cross sections from young **(A)** and old **(B)** autopsy specimens show a marked reduction in the amount of cancellous bone in the older specimen. (Reprinted with permission from Nordin, B. E. C.: Metabolic Bone and Stone Disease, p. 17. Edinburgh & London, Churchill Livingstone, 1973.)

TENSION

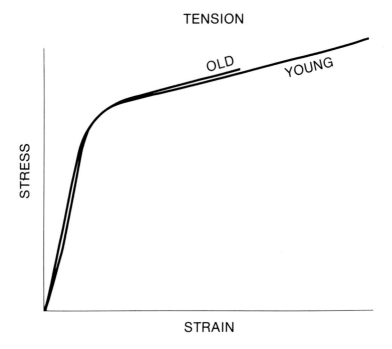

Fig. 1–48.   Stress-strain curves for samples of adult human tibiae of two widely differing ages tested in tension. (Adapted from Burstein et al.,1976.)

2. Bone is an anisotropic material, exhibiting different mechanical properties when loaded in different directions. Mature bone is strongest and stiffest in compression.
3. Bone increases in strength and stiffness and stores more energy with an increased speed of loading.
4. Muscle activity alters the in vivo stress pattern in bone.
5. Living bone fatigues when the frequency of loading precludes the remodeling necessary to prevent failure.
6. Bone remodels in response to the mechanical demands placed upon it; that is, bone is laid down when needed and resorbed where not needed (Wolff's law).

REFERENCES

Bonefield, W., and Li, C. H.: Anisotropy of nonelastic flow in bone. J. Appl. Physics, *38*:2450, 1967.
Burstein, A. H., Reilly, D. T., and Martens, M.: Aging of bone tissue: Mechanical properties. J. Bone Joint Surg., *58A*:82, 1976.
Burstein, A. H., Currey, J., Frankel, V. H., Heiple, K. G., Lunseth, P., and Vessely, J. C.: Bone strength. The effect of screw holes. J. Bone Joint Surg., *54A*:1143, 1972.
Carter, D. R.: Anisotropic analysis of strain rosette information from cortical bone. J. Biomech., *11*:199, 1978.

————: Personal communication, 1979.

Carter, D. R., and Hayes, W. C.: Bone compressive strength: The influence of density and strain rate. Science, *194*:1174, 1976.

————: Compact bone fatigue damage. A microscopic examination. Clin. Orthop., *127*:265, 1977a.

————: The compressive behavior of bone as a two-phase porous structure. J. Bone Joint Surg., *59A*:954, 1977b.

Dalén, N., and Olsson, K. E.: Bone mineral content and physical activity. Acta Orthop. Scand., *45*:170, 1974.

Frankel, V. H., and Burstein, A. H.: *Orthopaedic Biomechanics.* Philadelphia, Lea & Febiger, 1970.

Jenkins, D. P., and Cochran, T. H.: Osteoporosis: The dramatic effect of disuse of an extremity. Clin. Orthop., *64*:128, 1969.

Jones, H., Priest, J., Hayes, W., and Nagel, D.: Humeral hypertrophy in response to exercise. J. Bone Joint Surg., *59A*:204, 1977.

Kazarian, L. L., and Von Gierke, H. E.: Bone loss as a result of immobilization and chelation. Preliminary results in *Macaca mulatta*. Clin. Orthop., *65*:67, 1969.

Lanyon, L. E., Hampson, W. G. J., Goodship, A. E., and Shah, J. S.: Bone deformation recorded in vivo from strain gauges attached to the human tibial shaft. Acta Orthop. Scand., *46*:256, 1975.

Nilsson, B. E., and Westlin, N. E.: Bone density in athletes. Clin. Orthop., *77*:179, 1971.

Reilly, D., and Burstein, A.: The elastic and ultimate properties of compact bone tissue. J. Biomech., *8*:393, 1975.

Sammarco, J., Burstein, A., Davis, W., and Frankel, V.: The biomechanics of torsional fractures: The effect of loading on ultimate properties. J. Biomech., *4*:113, 1971.

SUGGESTED READING

Whole Bones

Asang, E.: Biomechanics of the human leg in alpine skiing. In *Biomechanics IV.* Edited by R. C. Nelson and C. A. Morehouse. Baltimore, University Park Press, 1974, pp. 236–242.

————: Applied biomechanics of the human leg. A basis for individual protection from skiing injuries. Orthop. Clin. North Am., *7*:95–103, 1976.

————: Experimental biomechanics of the human leg. A basis for interpreting typical skiing injury mechanisms. Orthop. Clin. North Am., *7*:63–73, 1976.

Burstein, A. H., and Frankel, V. H.: The viscoelastic properties of some biological materials. Ann. N.Y. Acad. Sci., *146*:158–165, 1968.

Burstein, A. H., Currey, J., Frankel, V. H., Heiple, K. G., Lunseth, P., and Vessely, J. C.: Bone strength. The effect of screw holes. J. Bone Joint Surg., *54A*:1143–1156, 1972.

Carter, D. R.: Anisotropic analysis of strain rosette information from cortical bone. J. Biomech., *11*:199–202, 1978.

Dalén, N., and Olsson, K.-E.: Bone mineral content and physical activity. Acta Orthop. Scand., *45*:170–174, 1974.

Evans, R. D. (ed.): *Studies on the Anatomy and Function of Bone and Joints.* New York, Springer-Verlag, 1966.

Frankel, V. H.: *The Femoral Neck: Function, Fracture Mechanisms, Internal Fixation.* Springfield, Charles C Thomas, 1960.

Frankel, V. H., and Burstein, A. H.: *Orthopaedic Biomechanics.* Philadelphia, Lea & Febiger, 1970.

————: Biomechanics of the locomotor system. In *Medical Engineering.* Edited by C. D. Ray. Chicago, Year Book Medical Publisher, 1974, pp. 505–516.

Frost, H. M.: *An Introduction to Biomechanics.* Springfield, Charles C Thomas, 1976.

Hakim, N. S., and King, A. I.: Programmed replication of *in situ* (whole body) loading conditions during *in vitro* (substructure) testing of a vertebral column segment. J. Biomech., *9*:629–632, 1976.

Inman, V. T.: Functional aspects of the abductor muscles of the hip. J. Bone Joint Surg., *29A*:607–619, 1947.

Jenkins, D. P., and Cochran, T. H.: Osteoporosis: The dramatic effect of disuse of an extremity. Clin. Orthop., *64*:128–134, 1969.

Jensen, J. S., Hansen, F. W., and Johansen, J.: Tibial shaft fractures. A comparison of conservative treatment and internal fixation with conventional plates or AO compression plates. Acta Orthop. Scand., *48*:204–212, 1977.

Jones, H., Priest, J., Hayes, W., and Nagel, D.: Humeral hypertrophy in response to exercise. J. Bone Joint Surg., *59A*:204–208, 1977.

Kazarian, L., and Graves, G. A.: Compressive strength characteristics of the human vertebral centrum. Spine, *2*:1–14, 1977.

Kazarian, L. E., and Von Gierke, H. E.: Bone loss as a result of immobilization and chelation. Preliminary results in *Macaca mulatta*. Clin. Orthop., *65*:67–75, 1969.

Lanyon, L. E., Hampson, W. G. J., Goodship, A. E., and Shah, J. S.: Bone deformation recorded in vivo from strain gauges attached to the human tibial shaft. Acta Orthop. Scand., *46*:256–268, 1975.

Murphy, E. F., and Burstein, A. H.: Physical properties of materials including solid mechanics. In *Atlas of Orthotics. Biomechanical Principles and Application.* St. Louis, C. V. Mosby Co., 1975, pp. 3–30.

Piziali, R. L., and Nagel, D. A.: Modeling of the human leg in ski injuries. Orthop. Clin. North Am., *7*:127–139, 1976.

Rosse, C., and Clawson, D. K.: *Introduction to the Musculoskeletal System.* New York, Harper & Row, 1970.

Rybicki, E. F., Simonen, F. A., and Weis, E. B.: On the mathematical analysis of stress in the human femur. J. Biomech., *5*:203–215, 1972.

Sammarco, G. J., Burstein, A. H., Davis, W. L., and Frankel, V. H.: The biomechanics of torsional fractures: The effect of loading on ultimate properties. J. Biomech., *4*:113–117, 1971.

Strömberg, L., and Dalén, N.: Experimental measurement of maximum torque capacity of long bones. Acta Orthop. Scand., *47*:257–263, 1976.

————: The influence of freezing on the maximum torque capacity of long bones. An experimental study on dogs. Acta Orthop. Scand., *47*:254–256, 1976.

Toridis, T. G.: Stress analysis of the femur. J. Biomech., *2*:163–174, 1969.

Cortical Bone

Bonefield, W., and Li, C. H.: Anisotropy of nonelastic flow in bone. J. Appl. Physics, *38*:2450–2455, 1967.

Burstein, A. H., Reilly, D. T., and Martens, M.: Aging of bone tissue: Mechanical properties. J. Bone Joint Surg., *58A*:82–86, 1976.

Burstein, A. H., Currey, J. D., Frankel, V. H., and Reilly, D. T.: The ultimate properties of bone tissue: The effects of yielding. J. Biomech., *5*:35–44, 1972.

Carter, D. R., and Hayes, W. C.: Compact bone fatigue damage. A microscopic examination. Clin. Orthop., *127*:265–274, 1977.

————: The compressive behavior of bone as a two-phase porous structure. J. Bone Joint Surg., *59A*:954–962, 1977.

Currey, J. D.: The mechanical properties of bone. Clin. Orthop., *73*:210–231, 1970.

Dalén, N., Hellström, L.-G., and Jacobson, B.: Bone mineral content and mechanical strength of the femoral neck. Acta Orthop. Scand., *47*:503–508, 1976.

Enneking, W.: *Principles of Musculoskeletal Pathology.* Gainesville, Florida, Storter Printing Company, 1970.

Evans, F. G.: *Mechanical Properties of Bone.* Springfield, Charles C Thomas, 1973.

Fredensborg, N., and Nilsson, B. E.: The bone mineral content and cortical thickness in young women with femoral neck fracture. Clin. Orthop., *124*:161–164, 1977.

Frost, H. M.: *The Laws of Bone Structure.* Springfield, Charles C Thomas, 1964.

Mueller, K. H., Trias, A., and Ray, R. D.: Bone density and composition. Age-related and pathological changes in water and mineral content. J. Bone Joint Surg., *48A*:140–148, 1966.

Nilsson, B. E., and Westlin, N. E.: Bone density in athletes. Clin. Orthop., *77*:179–182, 1971.

Pope, M. H., and Outwater, J. O.: The fracture characteristics of bone substance. J. Biomech., *5*:457–465, 1972.

Reilly, D. T., and Burstein, A. H.: The mechanical properties of cortical bone. J. Bone Joint Surg., *56A*:1001–1002, 1974.
———: The elastic and ultimate properties of compact bone tissue. J. Biomech., *8*:393–405, 1975.
Reilly, D. T., Burstein, A. H., and Frankel, V. H.: The elastic modulus for bone. J. Biomech., *7*:271–275, 1974.
Sedlin, E. D.: A rheological model for cortical bone. A study of the physical properties of human femoral samples. Acta Orthop. Scand., Suppl. *83*:1–87, 1965.
Viano, D., Helfenstein, U., Anliker, M., and Rüegsegger, P.: Elastic properties of cortical bone in female human femurs. J. Biomech., *9*:703–710, 1976.
Weaver, J. K.: The microscopic hardness of bone. J. Bone Joint Surg., *48A*:273–288, 1966.

Cancellous Bone

Behrens, J. C., Walker, P. S., and Shoji, H.: Variations in strength and structure of cancellous bone at the knee. J. Biomech., *7*:201–207, 1974.
Carter, D. R., and Hayes, W. C.: Bone compressive strength: The influence of density and strain rate. Science, *194*:1174–1176, 1976.
———: The compressive behavior of bone as a two-phase porous structure. J. Bone Joint Surg., *59A*:954–962, 1977.
Chung, S. M. K., Batterman, S. C., and Brighton, C. T.: Shear strength of the human femoral capital epiphyseal plate. J. Bone Joint Surg., *58A*:94–103, 1976.
Enneking, W.: *Principles of Musculoskeletal Pathology*. Gainesville, Florida, Storter Printing Company, 1970.
Evans, F. G.: *Mechanical Properties of Bone*. Springfield, Charles C Thomas, 1973.
Frankel, V. H., and Burstein, A. H.: Load capacity of tubular bone. In *Biomechanics and Related Bio-engineering Topics*. Edited by R. M. Kenedi. New York, Pergamon Press, 1965, pp. 381–396.
Frost, H. M.: *The Laws of Bone Structure*. Springfield, Charles C Thomas, 1964.
Galante, J., Rostoker, W., and Ray, R. D.: Physical properties of trabecular bone. Calcif. Tissue Res., *5*:236–246, 1970.
Hayes, W. C., and Carter, D. R.: Postyield behavior of subchondral trabecular bone. J. Biomed. Mater. Res., *7*:537–544, 1976.
Hayes, W. C., Boyle, D. J., and Velez, A.: Functional adaptation in the trabecular architecture of the human patella. Transactions of the 23rd Annual Meeting, Orthopaedic Research Society, *2*:114, 1977.
Lindahl, O.: Mechanical properties of dried defatted spongy bone. Acta Orthop. Scand., *47*:11–19, 1976.
Pope, M. H., and Outwater, J. O.: The fracture characteristics of bone substance. J. Biomech., *5*:457–465, 1972.
———: Mechanical properties of bone as a function of position and orientation. J. Biomech., *7*:61–66, 1974.
Weaver, J. K.: The microscopic hardness of bone. J. Bone Joint Surg., *48A*:273–288, 1966.
Weaver, J. K., and Chalmers, J.: Cancellous bone: Its strength and changes with aging and an evaluation of some methods for measuring its mineral content. I. Age changes in cancellous bone. J. Bone Joint Surg., *48A*:289–299, 1966.

Fatigue Fracture

Baker, J., Frankel, V. H., and Burstein, A. H.: Fatigue fractures: Biomechanical considerations. In Proceedings of the American Academy of Orthopaedic Surgeons. J. Bone Joint Surg., *54A*:1345–1346, 1972.
Burrows, H. J.: Fatigue infraction of the middle of the tibia in ballet dancers. J. Bone Joint Surg., *38B*:83–94, 1956.
Devas, M.: *Stress Fractures*. Edinburgh, London, and New York, Churchill Livingstone, 1975.
Frankel, V. H., and Hang, Y.-S.: Recent advances in the biomechanics of sport injuries. Acta Orthop. Scand., *46*:484–497, 1975.
Friedenberg, Z. B.: Fatigue fractures of the tibia. Clin. Orthop., *76*:111–115, 1971.

McBryde, A. M.: Stress fractures in athletes. J. Sports Med., 3:212–217, 1976.
Walter, N. E., and Wolf, M. D.: Stress fractures in young athletes. Am. J. Sports Med., 5:165–170, 1977.

Bone Remodeling

Abramson, A.: Bone disturbances in injuries to the spinal cord and cauda equina (paraplegia). Their prevention by ambulation. J. Bone Joint Surg., 30A:982–987, 1948.
Bartley, M. H., Arnold, J. S., Haslam, R. K., and Jee, W. S. S.: The relationship of bone strength and bone quantity in health, disease, and aging. J. Gerontol., 21:517–521, 1966.
Bassett, C. A. L.: Biologic significance of piezoelectricity. Calcif. Tissue Res., 1:252–272, 1968.
Chamay, A., and Tschantz, P.: Mechanical influences in bone remodeling. Experimental research on Wolff's Law. J. Biomech., 5:173–180, 1972.
Dalén, N., and Olsson, K.-E.: Bone mineral content and physical activity. Acta Orthop. Scand., 45:170–174, 1974.
Franke, J., Runge, H., Grau, P., Fengler, F., Wanka, C., and Remple, H.: Physical properties of fluorosis bone. Acta Orthop. Scand., 47:20–27, 1976.
Gjelsvik, A.: Bone remodeling and piezoelectricity—I. J. Biomech., 6:69–77, 1973.
Horal, J., Nachemson, A., and Scheller, S.: Clinical and radiological long term follow-up of vertebral fractures in children. Acta Orthop. Scand., 43:491–503, 1972.
Jenkins, D. P., and Cochran, T. H.: Osteoporosis: The dramatic effect of disuse of an extremity. Clin. Orthop., 64:128–134, 1969.
Jones, H. H., Priest, J. D., Hayes, W. C., Tichenor, C. C., and Nagel, D. A.: Humeral hypertrophy in response to exercise. J. Bone Joint Surg., 59A:204–208, 1977.
Schock, C. C., Noyes, F. R., Mathews, C. H. E., and Crouch, M. M.: The effect of activity on surface remodelling in the Rhesus monkey rib and femur. Transactions of the 23rd Annual Meeting, Orthopaedic Research Society, 2:36, 1977.

CHAPTER 2

Biomechanics of Joint Cartilage

Van C. Mow,
Vladimir Roth,
and
Cecil G. Armstrong

Articulating joints are the functional connections between different bones of the skeleton. In synovial, or freely moving, joints, the articulating bone ends are covered with a 1- to 5-mm dense white layer of connective tissue—the articular cartilage. Physiologically, articular cartilage is virtually an isolated tissue; it is devoid of blood and lymphatic supplies as well as nerves, and its cellular density is less than in most other tissues. The schematic zonal arrangement of these cells (chondrocytes) is shown in Figure 2–1.

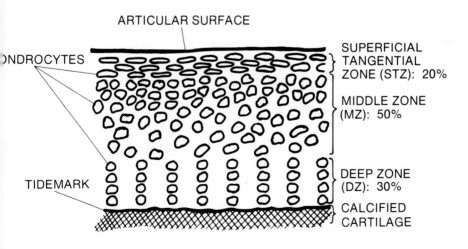

Fig. 2–1. Arrangement of chondrocytes throughout the depth of noncalcified cartilage. In the STZ, the chondrocytes are oblong with their long axes aligned parallel to the articular surface. In the MZ, the chondrocytes are "round" and are randomly distributed throughout. In the DZ, the chondrocytes are columnar in arrangement and are perpendicularly aligned to the tidemark. Chondrocytes are virtually isolated from the rest of the physiological system.

The main functions of this articular cartilage are (1) to spread the loads applied to the joint so that they are transmitted over a large area, i.e., so that the contact stresses are decreased; and (2) to allow relative movement of the opposing surfaces with minimum friction and wear. This chapter will describe how the biomechanical properties of the tissue, which are determined by the cartilage composition and structure, allow for the optimum performance of the above functions.

COMPOSITION

The solid matrix of articular cartilage, which accounts for 20 to 40% of the tissue's wet weight, is composed of collagen fibers (60%), an interfibrillar proteoglycan gel (40%) which has a high affinity for water, and cells, the chondrocytes (less than 2%). The remaining 60 to 80% of the tissue is water, most of which can be squeezed out under load. The relative juxtaposition of the main components of articular cartilage is schematically depicted in Figure 2–2 (Mow and Lai, 1979).

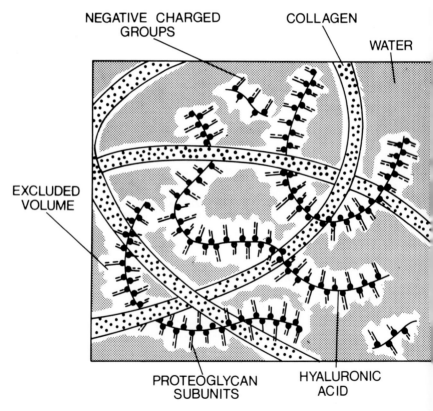

Fig. 2–2. A schematic depiction of proteoglycan, collagen, and water distribution throughout articular cartilage. (Adapted from Mow and Lai, 1979.)

The biomechanical behavior of this multiphasic structure is determined by the properties and distribution of the individual components and their interactions. The properties of the major solid constituents, i.e., collagen and proteoglycan, will be examined in greater detail first.

Collagen

Collagen is the most abundant protein in the animal body. It has a high level of structural organization, which results in optimum mechanical properties. The basic biological unit of collagen is a triple helix of polypeptide chains, called tropocollagen. Tropocollagen molecules are first arranged into a filament with a diameter of approximately 4 nanometers (1 nanometer equals 10^{-9} meter). These filaments can then be further polymerized into fibrils, fibers, and fiber bundles with much larger diameters (up to two microns in articular cartilage) (Clarke, 1971). It is thought (Muir, 1973) that proteoglycans play an important role in these larger aggregations, acting as bonding agents between fibrils and maintaining their ordered structure and physical properties (see below).

The most important mechanical properties of collagen are its tensile stiffness and strength. Tendons, 80% of whose dry weight may be collagen, have a tensile stiffness of one times 10^3 megapascals and a tensile strength of 50 megapascals. By way of comparison, steel has a tensile stiffness of approximately 200 times 10^3 megapascals and a tensile strength of approximately 700 megapascals. The tensile stiffness of aluminum is approximately 70 times 10^3 megapascals; its tensile strength is approximately 150 megapascals.

Collagen provides articular cartilage with a fibrous ultrastructure. This collagen network together with the hyperhydrated proteoglycans provides resistance to the stresses and strains of articulation. The uneven distribution of collagen throughout the cartilage thickness gives the tissue its layered character. Three separate structural zones have been identified in numerous light, electron, and scanning electron microscopic investigations of cartilage ultrastructure. A schematic description of the zonal arrangement of the fibrous collagen network proposed by Mow et al. (1974) is shown in Figure 2–3. The superficial tangential zone, representing between 10 and 20% of the total thickness, consists of densely packed fine fibrils arranged randomly within planes parallel to the articular surface (Weiss et al., 1968; Redler and Zimny, 1970). In the middle zone (40 to 60% of the total thickness), the interfibrillar spaces increase and the fibers are thought to be randomly distributed. Fibers in the most distant region (30% of the total thickness), the deep zone, are arranged in directions perpendicular to the cartilage-calcified cartilage interface, the tidemark, and are instrumental in anchoring the tissue to the underlying bone.

The nature of the collagen network, together with variations in its cross-link densities and/or collagen-proteoglycan interactions within planes

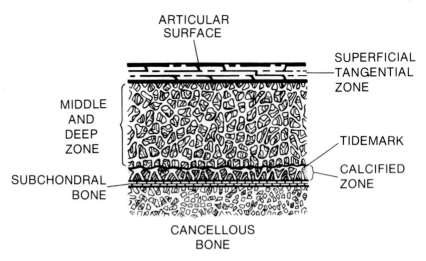

Fig. 2–3.   The ultrastructural arrangement of collagen fiber throughout the depth of articular cartilage. Note the correspondence between the chondrocyte distribution and this collagen fiber architecture. (Adapted from Mow et al., 1974.)

parallel to the articular surface, is thought to be a source of the directional dependence of the material properties, i.e., anisotropy. In tension, for example, this anisotropy is usually described with respect to the local split line direction at the articular surface. The split lines are elongated fissures produced by piercing the articular surface with a small awl (Hultkrantz, 1898). Figure 2–4 shows a typical split line pattern on the articular surface of a human femoral condyle. Its origin is probably related to the above-described directional variations in the material character of articular cartilage. However, the biophysical origin of these split lines remains unknown.

The Proteoglycan Gel

The other primary structural component of cartilage is a highly hydrated gel composed of a heterogeneous population of proteoglycan macromolecules and their polymers. The basic building blocks of proteoglycans are the glycosaminoglycans (GAGs). These GAGs are long flexible chains of repeating disaccharide units which are produced by assembling different combinations of these units. The differences between the sulphated GAGs of cartilage, i.e., keratan sulphate, chondroitin-4-sulphate, and chondroitin-6-sulphate, are small.

The first step in the formation of the extracellular ground substance constituents is the construction of a proteoglycan monomer. One end of each sulphated GAG is bound to a protein core 250 nanometers long. Approximately 30 GAGs can be accommodated per core. These proteo-

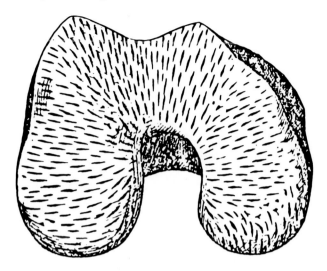

Fig. 2–4. Diagrammatical representation of a split line pattern on the surface of human femoral condyles. (Reprinted from Hultkrantz, 1898.)

glycan monomers are then attached at intervals ranging from 25 to 2,500 nanometers along the hyaluronic acid molecule to give a bottle-brush-type structure (Fig. 2–5) (Hardingham and Muir, 1974; Rosenberg, 1975). The length of this hyaluronic acid core ranges from 4 to 400 microns (one micron equals 10^{-6} meter), allowing room for over 150 proteoglycan monomers. The molecular weight of a structure of this size can be as great as two times 10^8 daltons.

There is believed to be a wide variation in the proteoglycan population in articular cartilage (McDevitt, 1973). Aggregates may vary in size and complexity from the giant molecule described above to the proteoglycan monomer. A special fraction of proteoglycans exist which are very closely associated with collagen and which probably bind the collagen fibrils together (Shepard and Mitchell, 1977). Similarly, the proteoglycan monomer, its protein core, and GAG side chains may all vary in size and composition.

The Structural Interaction Between the Components of Cartilage

Proteoglycans are strongly hydrophilic, occupying large solvent domains in the aqueous solution. This characteristic, accentuated by the presence of regular fixed negative charges on the GAG molecule, has several important consequences (Schubert and Hammerman, 1968; Maroudas, 1973): (1) a cloud of positive ions, e.g., Na^+, CA^{++}, is attracted to neutralize these fixed negative charges; (2) neighboring GAG chains are mutually repelled by their fixed charges and thus maintain a stiffly extended conformational state within the tissue; and (3) this concentrated solution of GAGs and small ions attempts to dilute itself by osmosis. Because the proteoglycans in cartilage are prevented from fully expanding by the collagen framework, there is an osmotic swelling pressure or "preload" on the matrix of roughly 0.35 megapascals. How the proteoglycans and the collagen fibers interact has

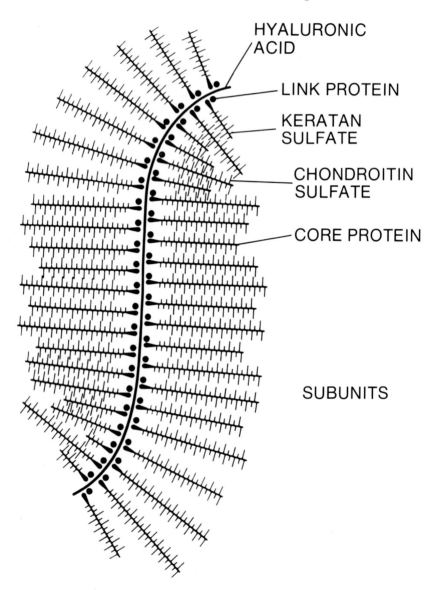

HYALURONIC ACID

LINK PROTEIN

KERATAN SULFATE

CHONDROITIN SULFATE

CORE PROTEIN

SUBUNITS

Fig. 2–5. Schematic depiction of a proteoglycan aggregate. The proteoglycan monomers are attached to the hyaluronic acid, forming a macromolecule of molecular weight of approximately two hundred million daltons. (Adapted from Rosenberg, 1975.)

not been fully determined. Simple entanglement is one form of interaction, and there is also a suggestion that proteoglycans can bind to the collagen molecules at certain sites (Serafini-Fracassini and Smith, 1974).

Even when no external stress is applied to the cartilage, the collagen fibers are in tension, constraining the swelling pressure of the proteoglycans. When a stress is applied to the cartilage surface, by a rigid indentor, for example, there is an instantaneous deformation, primarily caused by a change in the shape of the proteoglycan domains. Because of this externally applied stress, the internal pressure in the cartilage matrix now exceeds the swelling pressure, causing liquid to flow out of the tissue. This loss of water results in an increased concentration of the proteoglycan solution which in turn increases the swelling pressure. This process continues until the swelling pressure is eventually in equilibrium with the external stress. The rate of fluid movement during this process is controlled by the pressure difference and the permeability of the tissue. Tissue permeability is dependent upon the size and complexity of the molecular domains of both the proteoglycans and collagen. As GAG concentration increases, cartilage permeability generally decreases (Maroudas, 1973).

BIOMECHANICAL PROPERTIES OF ARTICULAR CARTILAGE

The biomechanical behavior of articular cartilage can only be understood in terms of a two-phase, i.e., biphasic, model. The two phases whose individual material properties and mutual interaction during load bearing determine the material behavior of the tissue are the solid organic matrix (collagen and proteoglycan) and the freely movable interstitial water. Thus, cartilage may be viewed as a fluid-filled porous medium (the analogy of a water-saturated sponge may be helpful). From the engineering point of view, materials of this type are rather difficult to analyze. The factors influencing the behavior of cartilage under load are the material characteristics of the solid matrix and its permeability.

Permeability

Permeability is a material parameter which represents the frictional resistance of the solid matrix of a porous material to the flow of fluid through it. The lower the permeability, the more resistance is offered to the fluid movement under the applied load. In comparison with the permeability of an ordinary sponge, that of healthy articular cartilage is miniscule. For normal bovine articular cartilage, the permeability has been determined to be 0.76 ± 0.30 times 10^{-14} meters^4 per newton second at a 95% statistical confidence level with an applied stress of 0.1 megapascal (Mow et al., submitted for publication). There are two primary mechanical means for fluid transport through porous media such as articular cartilage (Mow and Torzilli, 1975). These are schematically shown in Figure 2–6 A and B. The

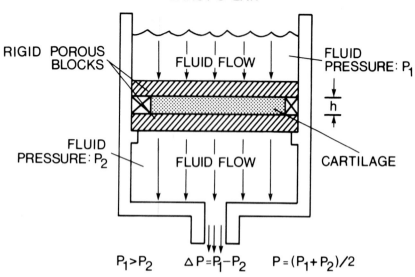

FLOW OF FLUID THROUGH CARTILAGE
DUE TO FLUID PRESSURE GRADIENT

A

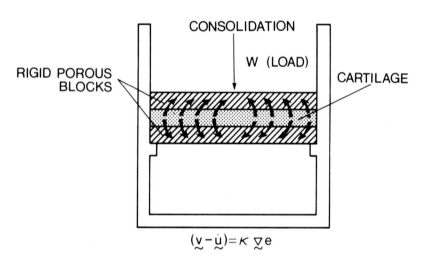

FLUID FLOW THROUGH CARTILAGE
DUE TO DEFORMATION

B

fluid can be forced through the porous solid matrix by application of a pressure gradient; i.e., the pressure on the top side of the cartilage sample is greater than the pressure on the bottom side, as in Figure 2–6A. Alternatively, if the fluid-saturated sample is placed under a rigid porous block and then compressed, as in Figure 2–6B, fluid flow will also occur. Flow in this case is induced by compressive deformation which decreases the solvent domain of the proteoglycan macromolecules and in turn results in a local increase in pressure. This pressure gradient is then the driving force causing exudation of the fluid from the tissue. Both of these mechanisms act simultaneously on articular cartilage during normal joint articulation. It has

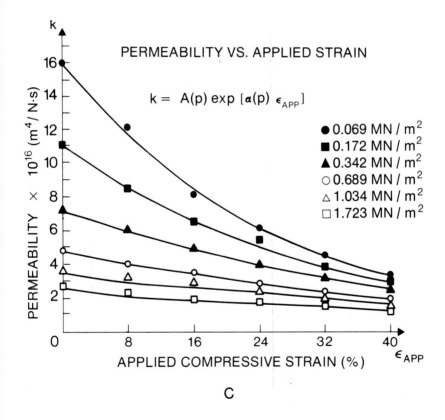

C

Fig. 2–6. **A.** Direction pressure gradient $(P_1 - P_2)/h$ being applied across the tissue sample. The permeability coefficient K = volume flow rate/[total area of perfusion $\times (P_1 - P_2)/h$]. (Adapted from Mow and Torzilli, 1975.)

   B. Exudation of the interstitial fluid subsequent to the application of load W, causing the consolidation of the solid matrix. (Adapted from Mow and Torzilli, 1975.)

   C. Decrease of permeability of articular cartilage as a function of applied compressive strain and applied pressure differential $(P_1 - P_2)$. The decrease may be described by the exponential law $k = A(p)\exp[\alpha(p)\epsilon_{APP}]$ where ϵ_{APP} is the applied compressive strain, p is the applied pressure, and $A(p)$ and $\alpha(p)$ are experimentally determined material parameters. (Adapted from Mow and Lai, 1979.)

been shown experimentally by Mansour and Mow (1976) that the permeability of healthy articular cartilage decreases dramatically with both increased pressure and deformation (Fig. 2–6C). Thus, articular cartilage has a mechanical feedback regulatory mechanism to prevent total removal of the interstitial fluid. This biomechanical regulatory system has profound implications for normal tissue nutritional requirements, joint lubrication, load-carrying capacity, and wear of the tissue.

In general, during pathological conditions, the continuity of the collagen-proteoglycan solid matrix becomes disrupted by either or both mechanical stresses or the biochemical effects of abnormal enzymatic actions. Thus, for example, the permeability of osteoarthritic tissue will be greater (due to defects in the collagen fiber network and a loss of the proteoglycan macromolecules) than that of normal tissue.

Rate Dependency of the Material Behavior of Articular Cartilage

Because of the high resistance of articular cartilage to fluid flow, i.e., low permeability, its material behavior is very much dependent on the rate at which the load is applied and removed. Thus, in a rapid loading and unloading situation, when there is not time for fluid to be squeezed out (for example, during jumping), the tissue will behave more or less like an elastic, single-phase material, deforming instantaneously upon loading and recovering instantaneously upon unloading. However, if the load is applied slowly or kept constant on the tissue, as for example during prolonged standing, the tissue deformation will continue to increase in time as the fluid is squeezed out. When unloaded, the tissue will recover its original dimensions provided it has enough fluid for a sufficient amount of time. The two types of material behavior described above are (1) time independent or elastic (recoverable) material behavior and (2) time dependent or viscoelastic (also recoverable) material behavior.

Behavior of Articular Cartilage Under Uniaxial Tension

The inhomogeneous-layered material character of articular cartilage described previously is reflected in its tensile behavior. To obtain tensile samples from the tissue, a full thickness sample is microtomed in planes parallel to its layered structure. Dumbbell-shaped tensile specimens are cut in these planes at specified angles (0, 45, and 90 degrees) with respect to the local split line direction at the articular surface. Specimens are then pulled to failure at low speed (0.5 cm per minute) in order to eliminate strain-rate dependent effects due to the viscoelastic character of the tissue.

A typical stress-strain curve for a tensile specimen of articular cartilage may be described by an exponential function (Fig. 2–7). Therefore, this type of material cannot be described by a single Young's modulus as can a linearly elastic material.

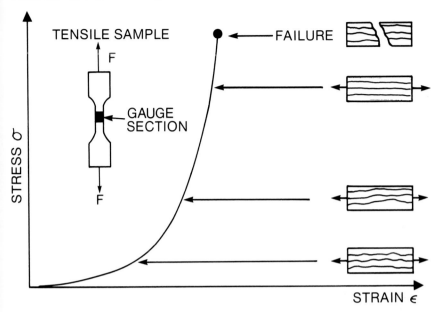

Fig. 2-7. Typical tensile stress-strain curve for articular cartilage strips. Usually the specimens taken parallel to the split lines are stiffer and stronger than those taken perpendicular to them.

The initial, smaller slope portion of the curve in Figure 2-7 is attributed to the alignment of the collagen structure in the direction of the applied tensile force. The final, nearly linear, steep portion of the curve represents the tensile stiffness of collagen itself. Again, any pathological or experimentally induced deviation from the composition or structure of healthy tissue, such as loss of proteoglycan, loss or degradation of the collagen, or decrease or increase of collagen cross-link density, will be reflected in an altered tensile response (Kempson et al., 1976).

Tensile stiffness and strength of normal adult articular cartilage were found to decrease with increasing distance from the articular surface (Kempson, 1973; Woo et al., 1976; Roth et al., 1978; Roth and Mow, in press). These results have led investigators to believe that the superficial zone, which is collagen rich and dense, acts like a tough, wear-resistant protective skin for the tissue.

Anisotropy of articular cartilage is reflected in the greater tensile strength and stiffness of samples cut parallel to the local split line direction than of samples cut perpendicular to it.

Creep Response of Articular Cartilage

Creep experiments are performed on viscoelastic materials as a way of determining one of their time-dependent material responses. A constant load is "instantaneously" applied to the sample and kept on for the duration of the experiment (for cartilage experiments, see Kuei et al., 1978). Under the applied load, the compressional deformation continuously increases, i.e., the sample "creeps," until it reaches a plateau or an asymptotic value (Fig. 2–8). Many viscoelastic theories can be used either to describe the behavior of this type of material or to obtain the appropriate material parameters.

Similar creep responses were obtained from experiments performed on cylindrical plugs of both articular cartilage and subchondral bone. However, since articular cartilage is biphasic, the articular surface of the specimen had to be loaded with a rigid, porous, free-draining filter to facilitate the flow of the fluid exuded from the sample. Thus, the creep process for articular cartilage is different from that of other viscoelastic materials. In the present case, fluid is initially expressed at a rapid rate (see Fig. 2–8), which is controlled by the instantaneous permeability of the tissue (see Fig. 2–6C). Eventually an equilibrium deformation is reached and no further exudation of the interstitial fluid occurs. At this stage, the swelling pressure and resistance to distortion of the collagen/proteoglycan gel is sufficient to support the applied load. The compressive strain and compressive stress at this equilibrium value can be used to determine the intrinsic modulus of the solid matrix. The value of this modulus is a reflection of the mechanical integrity for the tissue as a biomaterial. For normal bovine articular cartilage, this modulus has been determined to be 0.70 ± 0.064 megapascals at a 95%

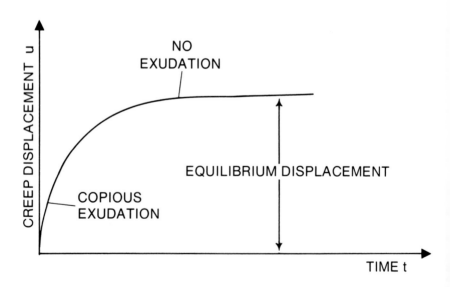

Fig. 2–8. Creep displacement of the articular surface loaded by a rigid porous filter. The continuous creep displacement is accompanied by copious amounts of fluid exudation. This phase of creep displacement is predominately governed by the permeability of the tissue. At equilibrium, no fluid flow occurs. The solid matrix supports all the applied load. (Adapted from Kuei et al.,1978.)

statistical confidence level and at an applied stress of 0.1 megapascals (Mow et al., submitted for publication). Deterioration of the matrix, e.g., loss of proteoglycan and disruption of the collagen fiber network, as a result of osteoarthrosis, for example, will result in a decrease of this modulus.

LUBRICATION

From the engineering point of view, there are only two fundamental types of lubrication: boundary and fluid film. Boundary lubrication depends on the chemical adsorption of a monolayer of lubricant molecules to the surfaces of the contacting solids (Bowden and Tabor, 1967). During relative motion, the surfaces of the bearing components are protected by lubricant molecules which slide over each other, preventing adhesion and abrasion of the naturally occurring surface asperities (Fig. 2–9). Boundary lubrication is essentially independent of the physical properties of either the lubricant (e.g., viscosity) or the contacting bodies (e.g., stiffness).

There is strong experimental evidence (Swann et al., 1974) that the synovial fluid in synovial joints can act, under some loading conditions, as a boundary lubricant for articular cartilage and further that this lubricating ability is not dependent on the viscosity of the synovial fluid. It is probable that this chemical adsorption of the synovial fluid to the articular surfaces is most important under severe loading conditions. Under less severe conditions, when the loads are lower and/or oscillate in magnitude, and the relative speed of the contacting surfaces is higher, it is quite possible that the joint may operate under the second type of lubrication mechanism, i.e., the fluid film mode.

In fluid film lubrication, a much thicker film of lubricant (compared with the molecular size of the lubricant in boundary lubrication) causes a

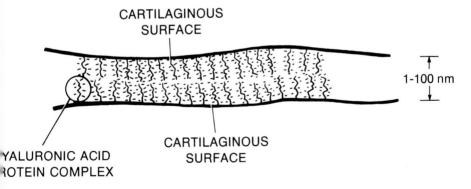

Fig. 2–9. Boundary lubrication of articular cartilage. Load is carried by a monolayer of macromolecules, probably a hyaluronic acid protein complex, adsorbed to the solid surfaces. These macromolecules appear to be effective in preventing cartilaginous wear (1 nm = 10^{-9} m). (Adapted from Armstrong and Mow, in press.)

relatively large separation of the two bearing surfaces. The load on the bearing is then supported by the pressure in this fluid film. In engineering bearings this fluid film thickness is usually less than 20 microns. When there is no relative sliding motion of the bearing surfaces, this pressure can be generated by an external pressure supply as in hydrostatic lubrication (Fig. 2–10A). If the surfaces are moving tangentially with respect to each other and if their shape is such that a converging wedge of fluid is formed, a lifting pressure is generated as the viscosity of the fluid causes it to be dragged into the gap between the surfaces (Fig. 2–10B). This mechanism is known as hydrodynamic lubrication. On the other hand, if the surfaces are moving perpendicularly toward each other, the fluid must be squeezed out from the gap between the two surfaces (Fig. 2–10C). Again, the viscosity and inertia of the fluid require that a pressure be generated in the fluid film to force out the lubricant. Obviously, the load cannot be supported indefinitely by this squeeze film lubrication process. Eventually, the film of fluid will be so thin that contact of the asperities on the two surfaces will occur. However, this mechanism is sufficient to carry high loads for short durations.

In squeeze film and hydrodynamic lubrication, the thickness, extent, and load-carrying capacity of the fluid film are determined only by the rheological properties of the lubricant (i.e., viscosity), the film geometry (i.e., the shape of the gap between the two surfaces), and the speed of the relative motion of the surfaces. However, if the bearing material is relatively soft, as is the case with articular cartilage, the pressure in the fluid film may cause substantial deformations of the contacting surfaces. These deformations beneficially alter the film geometry and contact area, causing a greater restriction to lubricant fluid escape and therefore a more substantial and longer lasting film (Fig. 2–11). This condition is known as elastohydrodynamic lubrication. Large increases in the load-carrying capacity of a bearing can be obtained in this way.

No surface, and especially that of articular cartilage, is perfectly smooth (Gardner and McGillivray, 1971). In a bearing, depending on operating conditions, situations may occur where the thickness of the fluid film generated is of the same order as the mean roughness of the bearing's surfaces. Contact between the peaks or asperities of the surfaces may then occur. When the load on the bearing is carried both by pressure in the partial fluid film and by boundary lubricated solid contacts, a mixed mode of lubrication occurs.

The special porous, fluid-saturated, permeable structure of articular cartilage allows yet another mode of lubrication. As the joint rotates, the loaded area sweeps over the articular surface, causing fluid to be exuded from the articular cartilage in front of and beneath the loaded contact area (Fig. 2–12). Once the area of peak stress has passed any given point, the fluid will start to be reabsorbed, ready for the next cycle of movement (Mansour and Mow, 1977). The volume of fluid expressed will not be large, but a film of fluid 10 microns thick will lubricate the cartilage surfaces

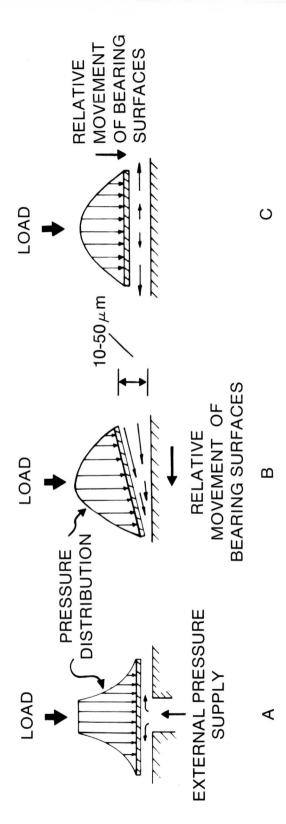

Fig. 2–10. Diagrams illustrating the load-carrying capacity of a fluid via **(A)** hydrostatic lubrication mechanism, **(B)** hydrodynamic lubrication mechanism, and **(C)** squeeze film lubrication mechanism. (Adapted from Armstrong and Mow, in press.)

75

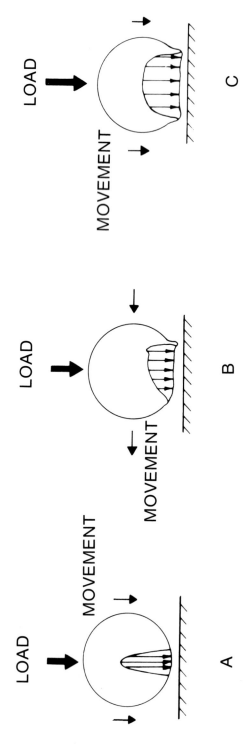

Fig. 2–11. Illustration of increased load-carrying capacity in elastohydrodynamic lubrication. Squeeze film lubrication of rigid surfaces (**A**), as compared with a sliding elastohydrodynamic lubrication process (**B**), and a squeeze film elastohydrodynamic lubrication process (**C**). (Adapted from Armstrong and Mow, in press.)

⇐ V VELOCITY OF THE CONTACT LOAD

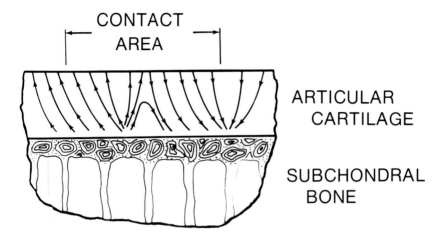

Fig. 2–12. Schematic representation of the velocity of the fluid flow through articular cartilage as it is subjected to a sliding load during joint function. The arrows within the cartilage indicate the direction of fluid flow. Fluid is exuded from the cartilage in front of and at the leading edge of the moving load and is imbibed at the trailing edge and behind the moving load. (Adapted from Mansour and Mow, 1976.)

extremely well. This forced circulation of fluid through the cartilage matrix also aids chondrocyte nutrition by bringing nutrients from the synovial fluid in the joint cavity into the cartilage cells.

Yet another mechanism by which the cartilage surfaces might be protected has been proposed by Walker et al. (1968, 1970). Their hypothesis, termed "boosted lubrication," is that it becomes progressively more difficult, as the two cartilage surfaces approach each other, for the large hyaluronic acid polymers in the synovial fluid to escape from the gap between the surface (Fig. 2–13). The water and small solute molecules can still escape either through the cartilage surface and/or laterally through the space between the two surfaces, but the large hyaluronic acid protein complexes may be physically too large to escape. This ultrafiltration process will eventually yield a concentrated gel of high viscosity which is probably less than 1 micron thick (Maroudas, 1973).

Summary of Lubrication Concepts

In comparison with engineering bearings, synovial joints are subjected to an enormous range of loading conditions; for example, in the human hip there will be (1) lightly loaded high-speed motions such as in the swing phase of walking or running (Paul, 1966/67), (2) impact loads of short

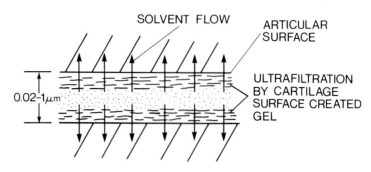

Fig. 2–13. Ultrafiltration of synovial fluid to form gel of concentrated hyaluronic acid solution between surfaces. (Adapted from Armstrong and Mow, in press.)

duration and large magnitude such as in jumping (Wright and Dowson, 1976) or in the heel strike of walking (Paul, 1976), and (3) fixed steady loads such as during prolonged standing. It is extremely unlikely that all of these varied demands upon this natural bearing can be satisfied by a single mode of lubrication. As yet, it is impossible to state definitely under which conditions a particular lubrication mechanism may operate. However, some general statements are possible. Again taking the human hip as an example:

1. Elastohydrodynamic fluid films of both the sliding and squeeze type probably play an important role in lubricating the joint. During the swing phase of walking, it is probable that quite a substantial film is generated. After heel strike, the film thickness will begin to decrease under high load by a squeeze film action. The second force peak during the walking cycle, just before the toe leaves the ground, acts in a different direction. The area under this second force peak was previously lightly loaded. It is possible that a healthy fluid film could exist at this area, thereby enabling further squeeze film action (Piotrowski, 1975).

2. Under high load and slow sliding speeds, the fluid film thickness will decrease. Under these conditions, the fluid squeezed out of the cartilage matrix could become the main contributor to the lubricating film. If the fluid film thins so much that contact of the surfaces occurs, even more fluid will be expressed to help support the load.

3. Under extreme conditions, such as an extended period of standing, the rate of fluid movement will decrease slowly, allowing more and more solid contacts between the asperities of the surface. However, the surface will probably still be protected, either by an ultrafiltrated synovial fluid gel or an adsorbed monolayer of boundary lubricant.

WEAR

Wear is the removal of material from solid surfaces by mechanical action. Like friction, wear can be divided into two components: interfacial wear due to the interaction of the bearing surfaces, and fatigue wear due to the deformation of the contacting bodies. If the bearing surfaces come into contact, interfacial wear can take place either by adhesion or by abrasion. Adhesive wear occurs if the junction caused by solid contact of the two surfaces is stronger than the underlying material. Fragments will then be torn off one surface and adhere to the other. Abrasion occurs when a soft material is scraped by a much harder one, which can be either an opposing surface or loose particles.

The low rates of wear observed in vitro suggest either that asperity contact of the two cartilage surfaces never or only rarely occurs, or that regeneration of the surface is possible (Radin and Paul, 1972). However, no evidence of ultrastructural repair of cartilage surfaces has ever been seen in aging or osteoarthrotic cartilage. Once ultrastructural defects and/or loss of mass of the surface occur, the surface layer of the cartilage becomes softer and more permeable. In this case, the resistance to fluid movement may be decreased, enabling the fluid to leak away through the cartilage from the fluid film separating the surfaces. This loss of fluid will increase the probability of solid contact of the surface asperities and further exacerbate the abrasion process.

Fatigue wear due to cyclically repeated deformation can take place even in well-lubricated bearings. Fatigue failure occurs because of the accumulation of microscopic damage in a material when it is repetitively stressed. Although the magnitude of the applied stresses may be much less than the material's ultimate strength, failure will eventually occur if they are applied often enough. In the synovial joint, rotation and sliding cause a given area of the articular surface to move in and out of the contact area. This process causes a repetitive stressing of the cartilage. Additionally, the total load on the joint varies cyclically during most physiological activities, again causing repetitive stressing. Due to the special mechanical properties of cartilage, there are several mechanisms by which fatigue damage might accumulate in the material. When cartilage is loaded, the load is supported by the collagen/proteoglycan matrix and by the resistance generated by fluid movement through it. Thus, repetitive loading and movement of the intact joint will cause repetitive stressing of the solid matrix and repetitive exudation and imbibition of the tissue's interstitial fluid.

Repetitive stressing on the collagen/proteoglycan matrix could cause disruption of (1) the collagen fibers, (2) the proteoglycan macromolecular network, or (3) the interface between the fibers and the interfibrillar matrix. One of the most popular hypotheses is that cartilage fatigue is due to a tensile failure of the collagen fiber framework (Freeman, 1975; Weightman, 1976). However, pronounced changes in the cartilage proteoglycan population with age and disease have also been observed (Inerot et al., 1978; Brandt and Palmoski, 1976). These changes might be considered to be part of the accumulated damage of the tissue.

Repetitive exudation and imbibition of the interstitial fluid may cause a "washing out" of the proteoglycans from the cartilage matrix near the articular surface (Torzilli and Mow, 1975). On the other hand, fluid movement away from areas of concentrated stress, such as those that occur in contact areas, allows the spreading or relaxation of these areas. This stress relaxation takes place quickly; the stress may decrease by 63% within 2 to 5 seconds (Mow, 1977). This phenomenon could well explain why Radin and Paul (1971) found impact loads to be so harmful to cartilage; the load was applied so quickly that there was no time for the fluid to move away from areas of high contact stress, and thus high stresses in the collagen/ proteoglycan matrix were produced.

Summary of Wear Concepts

The range of structural defects which can be observed in articular cartilage is large (Meachim and Fergie, 1975; Byers et al., 1977). For example, splitting of the cartilage surface is common. Vertical sections of cartilage exhibiting these lesions, known as fibrillation, show that they eventually extend through the full depth of the cartilage layer. On other occasions, the cartilage layer appears to be eroded rather than split. This erosion is known as smooth-surfaced destructive thinning.

It is unlikely that a single wear mechanism is responsible for all defects. At any given cartilage site, the history of stress at that point may be such that fatigue is the initiating failure mechanism. At another, the lubrication conditions may be so unfavorable that interfacial wear dominates the progression of joint failure. There is little experimental information on the type of defect produced by any given wear mechanism.

Once the cartilage microstructure is disrupted, any of the mechanical damage mechanisms mentioned above become possible (e.g., washing out of proteoglycans by violent fluid motions and impairment of articular cartilage's self-lubricating ability). These processes accelerate the interfacial wear rate and the fatigue rate of the already disrupted collagen/proteoglycan matrix.

BIOMECHANICS OF CARTILAGE DEGENERATION

Articular cartilage has a limited capacity for repair and regeneration, and, if the stresses to which it is subjected are large, total failure can occur quite quickly. It has been hypothesized that the progression of this failure is related to (1) the magnitude of the stresses experienced, (2) the total number of stress peaks experienced, and (3) the intrinsic molecular and microscopic structure of the collagen/proteoglycan matrix.

The magnitude of the stresses experienced by the cartilage is determined by both the total load on the joint and how that load is distributed over the contact area. The intensity of any concentrations of stress in the contact region will be of primary importance.

There are a large number of well-known conditions which cause excessive stress concentration and cartilage failure. Most are due to some type of incongruity of the joint resulting in an abnormally small contact area. Examples include osteoarthrosis subsequent to congenital acetabular dysplasia, slipped capital femoral epiphysis, or intra-articular fractures (Murray, 1965). Meniscectomy of the knee eliminates the load-spreading function of the menisci (Lutfi, 1975; Shrive et al., 1978), while ligament rupture allows excessive relative movement of the bone ends (Jacobsen, 1977), giving both an increased total load and an increased stress concentration as a consequence of the abnormal joint articulation.

These macroscopic stress concentrations have a further effect. High contact pressures between the articular surfaces decrease the probability of fluid film lubrication, since the pressure in the fluid must be sufficient to force the surfaces apart. Contact of the asperities on the solid surfaces then causes the microscopic stress concentrations which are responsible for the abrasion of material from those surfaces.

An increase in the frequency and magnitude of the total loads on the joint explains why the joints of people in certain occupations experience high incidences of degeneration. Examples include football players' knees, ballet dancers' ankles, and miners' knees. It has been suggested that some cases of osteoarthrosis may be due to deficiencies in the mechanisms which act to minimize peak joint forces. Examples of these mechanisms include the active processes of joint flexion and muscle lengthening, and the passive absorption of shocks by the subchondral bone (Radin, 1976).

Osteoarthrosis also occurs secondary to an insult to the intrinsic molecular and microscopic structure of the collagen/proteoglycan matrix. Examples of this phenomenon include the degeneration secondary to rheumatoid arthritis, hemorrhages into the joint space in hemophiliacs (Lee et al., 1974), various disorders of collagen metabolism, and possibly also the degradation of cartilage by proteolytic enzymes (Ali and Evans, 1973). The weakened cartilage structure is then destroyed by stresses of normal magnitude and frequency.

SUMMARY

1. The function of articular cartilage in diarthroidal joints is to transmit load and provide a smooth, wear-resistant bearing surface.
2. Important biomechanical properties of the articular cartilage are its permeability to interstitial fluid flow and the intrinsic moduli of the solid matrix.
3. The biomechanical properties of the cartilage depend upon the composition and macromolecular conformation of the proteoglycan aggregates and collagen fibrils.
4. In the joint, articular cartilage is "self-lubricating" under normal physiological conditions.

5. Damage to the tissue, whatever its cause, can disrupt the joint lubrication process and result in excessive wear.

ACKNOWLEDGMENTS

This work was sponsored by the National Science Foundation grant no. ENG-77-21247 and the National Institutes of Health grant no. AM19094-03. We also thank Ms. E. S. Miller and Mrs. M. Bungert, respectively, for their careful editing and typing of this manuscript.

REFERENCES

Ali, S. Y., and Evans, L.: Enzymic degradation of cartilage in osteoarthritis. Fed. Proc., *32*:1494, 1973.

Armstrong, C. G., and Mow, V. C.: Friction, lubrication and wear of synovial joints. In *Scientific Foundations of Orthopaedics and the Surgery of Trauma*. Edited by R. Owen, J. W. Goodfellow, and P. G. Bullough. London, William Heinemann Medical Books Ltd., in press.

Bowden, F. P., and Tabor, D.: *Friction and Lubrication*. London, Methuen, 1967.

Brandt, K. D., and Palmoski, M.: Organization of ground substance proteoglycans in normal and osteoarthritic knee cartilage. Arthritis Rheum., *19*:209, 1976.

Byers, P. D., Pringle, J., Oztop, F., Fernley, H. N., Brown, M. A., and Davison, W.: Observations on osteoarthrosis of the hip. Semin. Arthritis Rheum., *6*:277, 1977.

Clarke, I. C.: Surface characteristics of human articular cartilage—a scanning electron microscope study. J. Anat., *108*:23, 1971.

Freeman, M. A. R.: The fatigue of cartilage in the pathogenesis of osteoarthrosis. Acta Orthop. Scand., *46*:323, 1975.

Gardner, D. L., and McGillivray, D. C.: Living articular cartilage is not smooth. The structure of mammalian and avian joint surfaces demonstrated *in vivo* by immersion incident light microscopy. Ann. Rheum. Dis., *30*:3, 1971.

Hardingham, T. E., and Muir, H.: The function of hyaluronic acid in proteoglycan aggregation. In *Proceedings of the Symposium on Normal and Osteoarthrotic Articular Cartilage*. London, Institute of Orthopaedics, 1974, pp. 51–64.

Hultkrantz, J. W.: Über die Spaltrichtungen der Gelenkknorpel. Verhandl. D. Gesellsch., Jena:248, 1898.

Inerot, S., Heinegard, D., Audell, L., and Olsson, S.-E.: Articular-cartilage proteoglycans in aging and osteoarthritis. Biochem. J., *169*:143, 1978.

Jacobsen, K.: Osteoarthrosis following insufficiency of the cruciate ligament in man. Acta Orthop. Scand., *48*:520, 1977.

Kempson, G. E.: Mechanical properties of articular cartilage. In *Adult Articular Cartilage*. Edited by M. A. R. Freeman. London, Sir Isaac Pitman and Sons Ltd., 1973, pp. 171–227.

Kempson, G. E., Tuke, M. A., Dingle, J. T., Barrett, A. J., and Horsfield, P. H.: The effects of proteolytic enzymes on the mechanical properties of adult human articular cartilage. Biochim. Biophys. Acta, *428*:741, 1976.

Kuei, S. C., Mow, V. C., Lai, W. M., and Ancona, M. G.: Biphasic creep behavior of articular cartilage. Transactions of the 24th Annual Meeting, Orthopaedic Research Society, *3*:10, 1978.

Lee, P., Rooney, P. J., Sturrock, R. D., Kennedy, A. C., and Dick, W. C.: The etiology and pathogenesis of osteoarthrosis: A review. Semin. Arthritis Rheum., *3*:189, 1974.

Lutfi, A. M.: Morphological changes in articular cartilage after meniscectomy. J. Bone Joint Surg., *57B*:525, 1975.

Mansour, J. M., and Mow, V. C.: The permeability of articular cartilage under compressive strain and at high pressures. J. Bone Joint Surg., *58A*:509, 1976.

———: On the natural lubrication of synovial joints: Normal and degenerate. Transactions of the American Society of Mechanical Engineers, J. Lub. Tech., *99*:163, 1977.

Maroudas, A.: Physico-chemical properties of articular cartilage. In *Adult Articular Cartilage.* Edited by M. A. R. Freeman. London, Sir Isaac Pitman and Sons Ltd., 1973, pp. 131–170.

McDevitt, C. A.: Occasional survey. Biochemistry of articular cartilage. Ann. Rheum. Dis., *32*:364, 1973.

Meachim, G., and Fergie, I. A.: Morphological patterns of articular cartilage fibrillation. J. Pathol., *115*:231, 1975.

Mow, V. C.: Biphasic rheological properties of cartilage. In Proceedings of the 1977 Biomechanics and Biomaterials Symposium. Bull. Hosp. Joint Dis., *38*:121, 1977.

Mow, V. C., and Lai, W. M.: Mechanics of animal joints. Annu. Rev. Fluid Mech., *11*:247, 1979.

Mow, V. C., and Torzilli, P. A.: Fundamental fluid transport mechanisms through articular cartilage. Ann. Rheum. Dis., Suppl. *34*:82, 1975.

Mow, V. C., Lai, W. M., and Redler, I.: Some surface characteristics of articular cartilage. I. A scanning electron microscopy study and a theoretical mode for the dynamic interaction of synovial fluid and articular cartilage. J. Biomech., *7*:449, 1974.

Mow, V. C., Kuei, S. C., Lai, W. M., and Armstrong, C. G.: Biphasic creep and stress relaxation of articular cartilage in compression: Theory and experiments. Submitted to J. Biomech. Engng.

Muir, I. H. M.: Biochemistry. In *Adult Articular Cartilage.* Edited by M. A. R. Freeman. London, Sir Isaac Pitman and Sons Ltd., 1973, pp. 100–130.

Murray, R. O.: The aetiology of primary osteoarthritis of the hip. Br. J. Radiol., *38*:810, 1965.

Paul, J. P.: Forces transmitted by joints in the human body. Proceedings of the Institution of Mechanical Engineers, *181*, Part 3J: 8, 1966–67.

———: Force actions transmitted by joints in the human body. Proc. R. Soc. Lond., B, *192*:163, 1976.

Piotrowski, G.: Non-Newtonian lubrication of synovial joints. Ph.D. thesis, Case Western Reserve University, 1975.

Radin, E. L.: Aetiology of osteoarthrosis. Clin. Rheum. Dis., *2*:509, 1976.

Radin, E. L., and Paul, I. L.: Response of joints to impact loading. I. In vitro wear. Arthritis Rheum., *14*:356, 1971.

Radin, E. L., and Paul, I. L.: A consolidated concept of joint lubrication. J. Bone Joint Surg., *54A*:607, 1972.

Redler, I., and Zimny, M.: Scanning electron microscopy of normal and abnormal articular cartilage and synovium. J. Bone Joint Surg., *52A*:1395, 1970.

Rosenberg, L. C.: Structure of cartilage proteoglycans. In *Dynamics of Connective Tissue Macromolecules.* Edited by P. M. C. Burleigh and A. R. Poole. Amsterdam, North Holland Publisher, 1975, pp. 105–128.

Roth, V., and Mow, V. C.: The intrinsic tensile behavior of the matrix of bovine articular cartilage and its variation with age. J. Bone Joint Surg., in press.

Roth, V., Wirth, C. R., Mow, V. C., and Lai, W. M.: Variation of the tensile properties of articular cartilage with age. Transactions of the 24th Annual Meeting, Orthopaedic Research Society, *3*:9, 1978.

Schubert, M., and Hamerman, D.: *A Primer on Connective Tissue Biochemistry.* Philadelphia, Lea & Febiger, 1968.

Serafini-Fracassini, A., and Smith, J. W.: *The Structure and Biochemistry of Cartilage.* Edinburgh and London, Churchill Livingstone, 1974.

Shepard, N., and Mitchell, N.: The localization of articular cartilage proteoglycan by electron microscopy. Anat. Rec., *187*:463, 1977.

Shrive, N. G., O'Connor, J. J., and Goodfellow, J. W.: Load-bearing in the knee joint. Clin. Orthop., *131*:279, 1978.

Swann, D. A., Radin, E. L., Nazimiec, M., Weisser, P. A., Curran, N., and Lewinnek, G.: Role of hyaluronic acid in joint lubrication. Ann. Rheum. Dis., *33*:318, 1974.

Torzilli, P. A., and Mow, V. C.: Mathematical model for fluid transport through articular cartilage during function. 1975 Biomechanics Symposium, American Society of Mechanical Engineers, Applied Mechanics Division, *10*:109, 1975.

Walker, P. S., Dowson, D., Longfield, M. D., and Wright, V.: "Boosted lubrication" in synovial joints by fluid entrapment and enrichment. Ann. Rheum. Dis., *27*:512, 1968.

Walker, P. S., Unsworth, A., Dowson, D., Sikorski, J., and Wright, V.: Mode of aggregation of hyaluronic acid protein complex on the surface of articular cartilage. Ann. Rheum. Dis., *29*:591, 1970.

Weightman, B.: Tensile fatigue of human articular cartilage. J. Biomech., 9:193, 1976.

Weiss, C., Rosenberg, L., and Helfet, A. J.: An ultrastructural study of normal young adult human articular cartilage. J. Bone Joint Surg., 50A:663, 1968.

Woo, S. L-Y., Akeson, W. H., and Jemmott, G. F.: Measurements of nonhomogeneous directional mechanical properties of articular cartilage in tension. J. Biomech., 9:785, 1976.

Wright, V., and Dowson, D.: Lubrication and cartilage. J. Anat., 121:107, 1976.

SUGGESTED READING

Ali, S. Y., and Evans, L.: Enzymic degradation of cartilage in osteoarthritis. Fed. Proc., 32:1494–1498, 1973.

Armstrong, C. G., and Mow, V. C.: Friction, lubrication and wear of synovial joints. In *Scientific Foundations of Orthopaedics and the Surgery of Trauma.* Edited by R. Owen, J. W. Goodfellow, and P. G. Bullough. London, William Heinemann Medical Books Ltd., in press.

Armstrong, C. G., Bahrani, A. S., and Gardner, D. L.: *In vitro* measurement of articular cartilage deformations in the intact human hip under load. J. Bone Joint Surg., 61A:744–755, 1979.

Bollet, A. J.: Current comment: An essay on the biology of osteoarthritis. Arthritis Rheum., 12:152–163, 1969.

Bowden, F. P., and Tabor, D.: *Friction and Lubrication.* London, Methuen, 1967.

Brandt, K. D., and Palmoski, M.: Organization of ground substance proteoglycans in normal and osteoarthritic knee cartilage. Arthritis Rheum., 19:209–215, 1976.

Byers, P. D., Pringle, J., Oztop, F., Fernley, H. N., Brown, M. A., and Davison, W.: Observations on osteoarthrosis of the hip. Semin. Arthritis Rheum., 6:277–303, 1977.

Clarke, I. C.: Articular cartilage: A review and scanning electron microscope study. I. The interterritorial fibrillar architecture. J. Bone Joint Surg., 53B:732–750, 1971.

Clarke, I. C.: Surface characteristics of human articular cartilage—A scanning electron microscope study. J. Anat., 108:23–30, 1971.

Freeman, M. A. R.: The fatigue of cartilage in the pathogenesis of osteoarthrosis. Acta Orthop. Scand., 46:323–328, 1975.

Gardner, D. L., and McGillivray, D. C.: Living articular cartilage is not smooth. The structure of mammalian and avian joint surfaces demonstrated *in vivo* by immersion incident light microscopy. Ann. Rheum. Dis., 30:3–9, 1971.

Hardingham, T. E., and Muir, H.: The function of hyaluronic acid in proteoglycan aggregation. In *Proceedings of the Symposium on Normal and Osteoarthrotic Articular Cartilage.* London, Institute of Orthopaedics, 1974, pp. 51–64.

Hascall, V. C., and Sajdera, S. W.: Physical properties and polydispersity of proteoglycan from bovine nasal cartilage. J. Biol. Chem., 245:4920–4930, 1970.

Hultkrantz, J. W.: Über die Spaltrichtungen der Gelenkknorpel. Verhandl. D. Gesellsch., Jena:248–256, 1898.

Hayes, W. C., and Bodine, A. J.: Flow-independent viscoelastic properties of articular cartilage matrix. J. Biomech., 11:407–419, 1978.

Inerot, S., Heinegard, D., Audell, L., and Olsson, S.-E.: Articular-cartilage proteoglycans in aging and osteoarthritis. Biochem. J., 169:143–156, 1978.

Jacobsen, K.: Osteoarthrosis following insufficiency of the cruciate ligaments in man. Acta Orthop. Scand., 48:520–526, 1977.

Kempson, G. E.: Mechanical properties of articular cartilage. In *Adult Articular Cartilage.* Edited by M. A. R. Freeman. London, Sir Isaac Pitman and Sons Ltd., 1973, pp. 171–227.

Kempson, G. E., Tuke, M. A., Dingle, J. T., Barrett, A. J., and Horsfield, P. H.: The effects of proteolytic enzymes on the mechanical properties of adult human articular cartilage. Biochim. Biophys. Acta, 428:741–760, 1976.

Kuei, S. C., Mow, V. C., Lai, W. M., and Ancona, M. G.: Biphasic creep behavior of articular cartilage. Transactions of the 24th Annual Meeting, Orthopaedic Research Society, 3:10, 1978.

Lee, P., Rooney, P. J., Sturrock, R. D., Kennedy, A. C., and Dick, W. C.: The etiology and pathogenesis of osteoarthrosis: A review. Semin. Arthritis Rheum., 3:189–218, 1974.

Lutfi, A. M.: Morphological changes in articular cartilage after meniscectomy. J. Bone Joint Surg., 57B:525–528, 1975.

Mankin, H. J.: The structure, chemistry, and metabolism of articular cartilage. Bull. Rheum. Dis., *17*:447–452, 1967.

Mansour, J. M., and Mow, V. C.: The permeability of articular cartilage under compressive strain and at high pressures. J. Bone Joint Surg., *58A*:509–516, 1976.

———: On the natural lubrication of synovial joints: Normal and degenerate. Transactions of the American Society of Mechanical Engineers, J. Lub. Tech., *99*:163–173, 1977.

Maroudas, A.: Physico-chemical properties of articular cartilage. In *Adult Articular Cartilage*. Edited by M. A. R. Freeman. London, Sir Isaac Pitman and Sons Ltd., 1973, pp. 131–170.

McDevitt, C. A.: Occasional survey. Biochemistry of articular cartilage. Nature of proteoglycans and collagen of articular cartilage and their role in ageing and in osteoarthrosis. Ann. Rheum. Dis., *32*:364–378, 1973.

Meachim, G., and Fergie, I. A.: Morphological patterns of articular cartilage fibrillation. J. Pathol., *115*:231–240, 1975.

Mow, V. C.: Biphasic rheological properties of cartilage. In Proceedings of the 1977 Biomechanics and Biomaterials Symposium. Bull. Hosp. Joint Dis., *38*:121–124, 1977.

Mow, V. C., and Lai, W. M.: Mechanics of animal joints. Annu. Rev. Fluid Mech., *11*:247–288, 1979.

Mow, V. C., and Torzilli, P. A.: Fundamental fluid transport mechanisms through articular cartilage. Ann. Rheum. Dis., Suppl. *34*:82–84, 1975.

Mow, V. C., Lai, W. M., and Redler, I.: Some surface characteristics of articular cartilage. I. A scanning electron microscopy study and a theoretical mode for the dynamic interaction of synovial fluid and articular cartilage. J. Biomech., *7*:449–456, 1974.

Mow, V. C., Kuei, S. C., Lai, W. M., and Armstrong, C. G.: Biphasic creep and stress relaxation of articular cartilage in compression: Theory and experiments. Submitted to J. Biomech. Engng.

Muir, I. H. M.: Biochemistry. In *Adult Articular Cartilage*. Edited by M. A. R. Freeman. London, Sir Isaac Pitman and Sons Ltd., 1973, pp. 100–130.

Murray, R. O.: The aetiology of primary osteoarthritis of the hip. Br. J. Radiol., *38*:810–824, 1965.

Paul, J. P.: Forces transmitted by joints in the human body. Proceedings of the Institution of Mechanical Engineers, *181*, Part 3J:8–15, 1966–67.

———: Force actions transmitted by joints in the human body. Proc. R. Soc. Lond., B, *192*:163–172, 1976.

Piotrowski, G.: Non-Newtonian lubrication of synovial joints. Ph.D. thesis, Case Western Reserve University, 1975.

Radin, E. L.: Aetiology of osteoarthrosis. Clin. Rheum. Dis., *2*:509–522, 1976.

Radin, E. L., and Paul, I. L.: Response of joints to impact loading. I. In vitro wear. Arthritis Rheum., *14*:356–362, 1971.

Radin, E. L., and Paul, I. L.: A consolidated concept of joint lubrication. J. Bone Joint Surg., *54A*:607–616, 1972.

Redler, I., and Zimny, M.: Scanning electron microscopy of normal and abnormal articular cartilage and synovium. J. Bone Joint Surg., *52A*:1395–1404, 1970.

Rosenberg, L. C.: Structure of cartilage proteoglycans. In *Dynamics of Connective Tissue Macromolecules*. Edited by P. M. C. Burleigh and A. R. Poole. Amsterdam, North Holland Publisher, 1975, pp. 105–128.

Roth, V., and Mow, V. C.: The intrinsic tensile behavior of the matrix of bovine articular cartilage and its variation with age. J. Bone Joint Surg., in press.

Roth, V., Mow, V. C., and Grodzinsky, A. J.: Biophysical and electromechanical properties of articular cartilage. In *Skeletal Research*. Edited by D. J. Simmons and A. S. Kunin. New York, San Francisco, London, Academic Press, 1979, pp. 301–341.

Roth, V., Wirth, C. R., Mow, V. C., and Lai, W. M.: Variation of the tensile properties of articular cartilage with age. Transactions of the 24th Annual Meeting, Orthopaedic Research Society, *33*:9, 1978.

Schubert, M., and Hamerman, D.: *A Primer on Connective Tissue Biochemistry*. Philadelphia, Lea & Febiger, 1968.

Serafini-Fracassini, A., and Smith, J. W.: *The Structure and Biochemistry of Cartilage*. London and Edinburgh, Churchill Livingstone, 1974.

Shepard, N., and Mitchell, N.: The localization of articular cartilage proteoglycan by electron microscopy. Anat. Rec., *187*:463–476, 1977.

Shrive, N. G., O'Connor, J. J., and Goodfellow, J. W.: Load-bearing in the knee joint. Clin. Orthop., *131*:279–287, 1978.

Swann, D. A., Radin, E. L., Nazimiec, M., Weisser, P. A., Curran, N., and Lewinnek, G.: Role of hyaluronic acid in joint lubrication. Ann. Rheum. Dis., *33*:318–326, 1974.

Torzilli, P. A.: The lubrication of human joints: A review. In *Handbook of Engineering in Medicine and Biology.* Edited by D. G. Fleming and B. N. Feinberg. West Palm Beach, Florida, CRC Press Inc., 1976, pp. 225–251.

Torzilli, P. A., and Mow, V. C.: Mathematical model for fluid transport through articular cartilage during function. 1975 Biomechanics Symposium, American Society of Mechanical Engineers, Applied Mechanics Division, *10*:109–113, 1975. ·

Walker, P. S., Dowson, D., Longfield, M. D., and Wright, V.: "Boosted lubrication" in synovial joints by fluid entrapment and enrichment. Ann. Rheum. Dis., *27*:512–520, 1968.

Walker, P. S., Unsworth, A., Dowson, D., Sikorski, J., and Wright, V.: Mode of aggregation of hyaluronic acid protein complex on the surface of articular cartilage. Ann. Rheum. Dis., *29*:591–602, 1970.

Weightman, B.: Tensile fatigue of human articular cartilage. J. Biomech., *9*:193–200, 1976.

Weiss, C., Rosenberg, L., and Helfet, A. J.: An ultrastructural study of normal young adult human articular cartilage. J. Bone Joint Surg., *50A*:663–674, 1968.

Woo, S. L-Y., Akeson, W. H., and Jemmott, G. F.: Measurements of nonhomogeneous directional mechanical properties of articular cartilage in tension. J. Biomech., *9*:785–791, 1976.

Wright, V., and Dowson, D.: Lubrication and cartilage. J. Anat., *121*:107–118, 1976.

CHAPTER 3

BIOMECHANICS OF COLLAGENOUS TISSUES

Margareta Nordin
and
Victor H. Frankel

The collagenous tissues surrounding the skeletal system are the ligaments (including the joint capsules), the tendons, and the skin. These structures are passive; that is, they do not inherently produce active motion.

The collagenous tissues are composed primarily of three types of fibers: collagen fibers, elastic fibers, and reticulin fibers. The collagen fibers provide strength and stiffness to the tissue, the elastic fibers give the tissue its extensibility under load, and the reticulin fibers provide bulk. An additional component of collagenous tissues is the ground substance, a gelatinous material that reduces friction between the fibers.

During activity the ligaments and tendons are mainly loaded in tension. Joint motion produces tensile loads on the ligaments, whereas muscle contraction produces these loads on the tendons. The skin is loaded in a more complex manner, withstanding tensile, compressive, and shear loading.

This chapter will deal primarily with the structure and also the mechanical behavior and properties of two collagenous tissues: the ligaments, including the joint capsules, and the tendons. The structure of the skin will be described, but its mechanical behavior and properties will not be discussed.

MECHANICAL PROPERTIES OF COLLAGENOUS TISSUES

The behavior of the collagenous tissues under loading is influenced by three main factors:
1. The structural orientation of the fibers
2. The properties of the collagen and elastic fibers
3. The proportion between the collagen and elastic fibers

Structural Orientation of Fibers

The structural orientation of the fibers differs in the three collagenous tissues and is suited to the function of each tissue (Fig. 3–1). The fibers of the

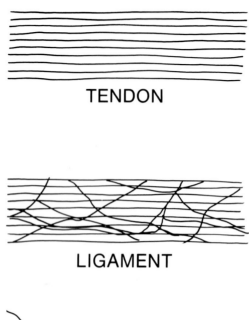

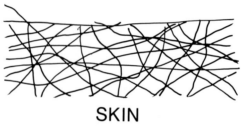

Fig. 3–1. Schematic diagram of the structural orientation of the fibers of tendon, ligament, and skin.

tendons have an almost completely parallel alignment, which makes the tendons well suited for withstanding high tensile loads. The fibers of the ligaments, including the joint capsules, have a less consistent structural orientation which varies in different ligaments depending on the function of the ligament (Kennedy et al., 1976). Although most ligament fibers are nearly parallel, some have a nonparallel orientation. The fibers of the skin have no predominant orientation but are intermeshed. This structural arrangement gives the skin its extensibility in all directions.

Fig. 3–2. Scanning electron micrographs of unloaded (relaxed) and loaded collagen fibers of human knee ligaments (10,000×).
   A. The unloaded collagen fibers have a wavy configuration.
   B. The loaded collagen fibers have straightened out. (Reprinted with permission from Kennedy, J. C., et al.: Tension studies of human knee ligaments. Yield point, ultimate failure, and disruption of the cruciate and tibial collateral ligaments. J. Bone Joint Surg., 58A:350–355, 1976.)

A

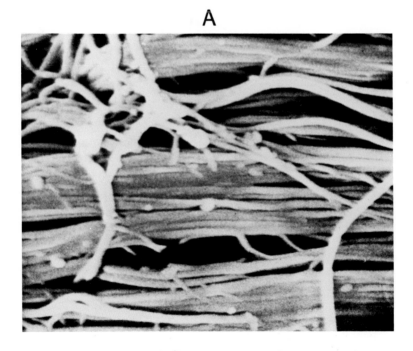

B

The fibers of the collagenous tissues have a different configuration when loaded and unloaded. When the fibers are relaxed (unloaded), they have a wavy configuration (Fig. 3–2A). When loads are imposed, low loads are sustained until the fibers oriented in the direction of loading straighten out (Fig. 3–2B). At this point the straightened fibers sustain loads in the physiologic range.

The differences in the alignment of the fibers in tendon, ligament, and skin produce differences in the mechanical properties of these tissues. When tendons are loaded in tension, all the fibers straighten out because of their parallel alignment. Of the three collagenous tissues, tendons are able to bear the highest tensile loads because all the fibers are aligned parallel to the direction of loading.

In tensile loading of the ligaments, where the fibers are not as aligned, only the fibers oriented in the direction of the principal load straighten out completely at first and sustain maximum loads. Those not oriented parallel to the direction of loading are subjected to lower loads until they have straightened out.

In tensile loading of skin, where the fibers are even more randomly oriented than in ligaments, markedly fewer fibers are oriented in the direction of loading and sustain maximum loads. Thus, the skin is weaker in tension than tendon and ligament.

Properties of Collagen and Elastic Fibers

The main components of the collagenous tissues are collagen fibers and elastic fibers, which together constitute approximately 90% of the tissues. These two fiber types behave differently under loading because collagen fibers are a ductile-like material and elastic fibers are a brittle-like material (Grood, 1978).

The nearly ductile and brittle behavior of the two fiber types was demonstrated in tensile tests to failure of parallel collagen fibers from human tendon (Abrahams, 1967) and parallel elastic fibers from human muscle (Carton et al., 1962) (Fig. 3–3). During tensile testing the collagen fibers elongated slightly as loading was begun, but rapidly became stiff with increased load until a yield point was reached (Fig. 3–3A). Nonelastic deformation then took place until ultimate failure. Deformation to failure ranged from 6 to 8%. The elastic fibers displayed a great elongation (more than two times their original length) when low loads were imposed. With increased loads, however, they suddenly became stiff and ruptured abruptly without deformation (Fig. 3–3B).

Collagen fiber bundles are relatively strong and tolerate about one-half the stress tolerated by cortical bone in tension. Elastic fiber bundles are relatively weak, tolerating only about one-tenth the stress tolerated by cortical bone in tension.

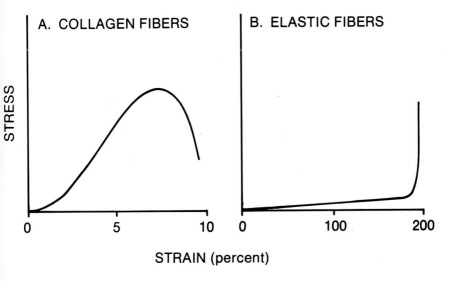

Fig. 3–3. Stress-strain curves for collagen and elastic fiber bundles tested in tension to failure.
A. The collagen fibers elongated slightly when low loads were applied until the wavy fibers had straightened out. At this point the collagen fiber bundles rapidly became stiff until the yield point was reached. Nonelastic deformation then took place until ultimate failure. (Adapted from Abrahams, 1967.)
B. The elastic fibers elongated greatly when low loads were imposed. They suddenly became stiff when the ultimate failure point was reached and ruptured abruptly without deforming. (Adapted from Carton et al., 1962.)

Proportion Between Collagen and Elastic Fibers

The proportion between the elastic and collagen fibers in the collagenous tissues varies according to the function each tissue performs and influences the mechanical behavior of the tissue. The primary function of a tendon is to transmit muscle forces to bone or fascia. This tissue consists almost entirely of collagen fibers, and therefore its behavior under tensile loading is almost identical to that of the isolated collagen bundles shown in Figure 3–3A.

The primary function of a ligament, including the joint capsule, is to stabilize a joint during motion and to prevent excessive motion. Like the tendons, most ligaments in the body consist mainly of collagen fibers. Two ligaments in the spine, however, the ligamentum nuchae and ligamentum flavum, are composed of two-thirds elastic fibers and display almost completely elastic behavior (Fielding et al., 1976; Nachemson and Evans, 1968). It appears that these ligaments have a specialized function, which might be to protect the nerve roots from mechanical impingement, to prestress the discs, and to provide intrinsic stability to the spine. Figure 3–4 demonstrates the difference in the mechanical behavior of the two ligaments experimentally tested in tension to failure: one, the anterior cruciate ligament in the knee, with a high percentage of collagen fibers (90%) (Noyes, 1977), and the other, the ligamentum flavum, with a high percentage of elastic fibers (60 to 70%) (Nachemson and Evans, 1968).

The load-deformation curve for the anterior cruciate ligament (Fig. 3–4A) is typical for a collagenous tissue with a high percentage of collagen fibers, such as tendon and most ligaments. This curve has five distinct regions:

1. Primary region. In this region elongation of the tissue occurs with a low load as the wavy collagen fibers straighten out (the "toe" portion of the curve).

2. Secondary region. The fibers oriented in the direction of loading straighten out completely, and the stiffness of the tissue increases rapidly. Deformation of the tissue begins and has a linear relationship with load. Microfailure of the collagen fibers also begins.

3. Third region. Progressive failure of the collagen fiber bundles takes place after a strain of 6 to 8% has been exceeded (yield point), but the tissue still has a normal gross appearance.

4. Fourth region. Major failure of the tissue occurs as the maximum load that the tissue can sustain is reached. Tensile failure of the collagen fibers and shear failure between the fibers are the main failure mechanisms.

5. Fifth region. Complete failure of the ligament takes place at about 6 to 8% elongation. The tissue is not able to support loads but is still in continuity.

The load-deformation curve for the ligamentum flavum (Fig. 3–4B), typical for this ligament and the ligamentum nuchae, is entirely different from that for the anterior cruciate ligament. In this ligament the elongation

---

Fig. 3–4. Mechanical behavior of two ligaments experimentally tested in tension to failure: one with a high percentage of collagen fibers and the other with a high percentage of elastic fibers.

    A. Load-elongation curve for a human anterior cruciate ligament (90% collagen fibers) tested in tension to failure. (Adapted from Noyes, 1977.)

      1. In the primary region (the "toe" portion), elongation occurred with a small increase in load as the wavy collagen fibers straightened out.

      2. In the secondary region the fibers oriented in the direction of loading became fully straightened out and the stiffness of the tissue increased rapidly. Deformation of the tissue began and had a linear relationship with load. Microfailure of the collagen fibers also began.

      3. In the third region progressive failure of the collagen fiber bundles took place after a strain of 6 to 8% had been exceeded (yield point), but the tissue still had a normal gross appearance.

      4. In the fourth region major failure of the tissue occurred as the maximum load was reached. Tensile failure of the collagen fibers and shear failure between the fibers were the main failure mechanisms.

      5. In the fifth region complete failure of the ligament took place at about 6 to 8% elongation. Although the tissue was still in continuity, it was not able to support loads.

    B. Load-elongation curve for a human ligamentum flavum (60 to 70% elastic fibers) tested in tension to failure. At 70% elongation the ligament failed abruptly. (Adapted from Nachemson and Evans, 1968.)

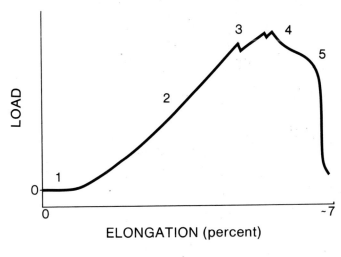

A

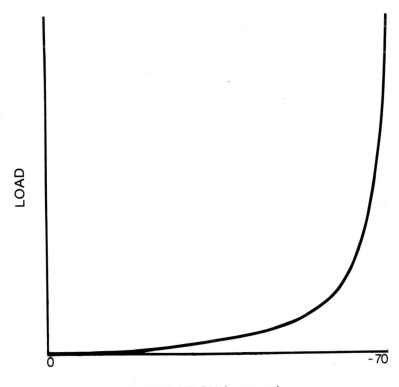

B

reached 50% before the stiffness increased. At this point, however, the stiffness increased rapidly and the ligament failed abruptly.

LIGAMENTS

The function of the ligaments, including the joint capsules, is to stabilize the joints, to guide joint motion, and to prevent excessive motion.

Two main factors determine the strength of a ligament under loading: the size and shape of the ligament, and the speed of loading. The cross-sectional area of a ligament influences its strength. The greater the number of fibers that are oriented in the direction of loading, and the wider and thicker these fibers are, the stronger the ligament. Like bone, a ligament increases in strength and stiffness with an increased speed of loading. Kennedy et al. (1976) found an almost 50% increase in load to failure when the loading speed (deformation rate) was increased fourfold during tensile testing of ligaments of the knee joint.

Bone-Ligament-Bone Complex

The mechanical behavior of the ligaments has been considered in isolation. However, these structures must be considered as a link in a bone-ligament-bone complex. Cooper and Misol (1970) examined the insertion of ligaments in dog knees by light and electron microscopy and divided the insertion into four zones on the basis of its histologic appearance (see Fig. 3–11). At the end of the ligament (zone 1), the collagen fibers intermesh with fibrocartilage (zone 2). This fibrocartilage gradually becomes mineralized fibrocartilage (zone 3). The mineralized fibrocartilage then merges with cortical bone (zone 4). The stress concentration effect at the insertion of the ligament into the stiffer bone structure is decreased by the presence of the three progressively stiffer materials at the ligament-bone junction (zones 1, 2, and 3).

Joint Displacement During Ligament Failure

When a ligament is subjected to loading, microfailure takes place even before the yield point is reached. When the yield point is exceeded, the ligament begins to undergo gross failure, and simultaneously the joint begins to displace abnormally. Because failure of a ligament leads to a large joint displacement, damage can also occur to the surrounding structures such as the joint capsule and other ligaments. Noyes (1977) applied a clinical test, the anterior drawer test, to a cadaver knee up to the point of anterior cruciate ligament failure. Figure 3–5 depicts the progressive failure of this ligament and displacement of the joint. At maximum load the joint had displaced several millimeters. The ligament was still in continuity even though it had undergone extensive macro- and microfailure and extensive elongation. The

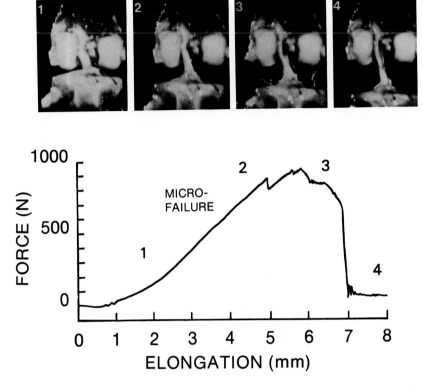

Fig. 3–5. Progressive failure of the anterior cruciate ligament from a cadaver knee tested in tension to failure at a physiological strain rate (Noyes, 1977). Almost 8 mm of joint displacement took place before the ligament reached complete failure. The force-elongation curve generated during this experiment is correlated with various degrees of joint displacement. (Courtesy of Frank R. Noyes, M.D., and Edward S. Grood, Ph.D.)

figure also shows the force-elongation curve generated during this experiment, which indicates where microfailure of the ligament begins.

The results of this in vitro test can be correlated with clinical findings. In Figure 3–6 the curve for the experimental study is divided into three regions. The first region corresponds to the amount of load placed on a ligament during tests of joint stability performed clinically. The second region corresponds to the amount of load placed on the ligament during physiological activity. The third region corresponds to the amount of load imposed on the ligament from the beginning of microfailure to complete rupture.

Ligament injuries fall into three categories depending on their severity. Injuries in the first category produce negligible clinical symptoms. Some pain is felt, but no joint instability can be detected clinically. However, microfracture of the collagen fibers may have occurred.

Injuries in the second category produce severe pain, and some joint instability is detected clinically. Progressive failure of the collagen fibers has

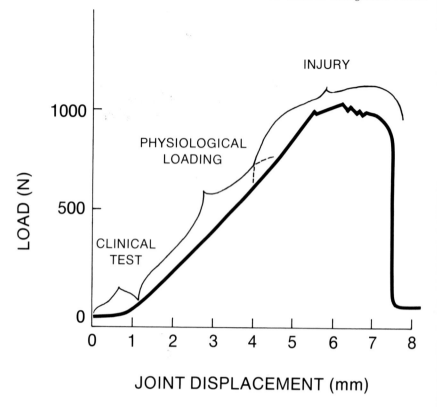

Fig. 3–6. The curve shown in Figure 3–5 has been converted to a load-displacement curve
and divided into three regions which correlate with clinical findings:
1. Amount of load imposed on the anterior cruciate ligament during the anterior drawer
   test
2. Amount of load placed on the ligament during physiological activity
3. Amount of load imposed on the ligament from partial injury to complete rupture

taken place, producing partial ligament rupture. The strength and stiffness of
the ligament may have decreased by 50% or more. Often the joint instability
produced by a partial rupture of a ligament is masked by muscle activity,
and thus the clinical test for joint stability is usually performed under
anesthesia.

Injuries in the third category produce severe pain during the course of
trauma with less pain after injury. Clinically the joint is found to be
completely unstable. Most collagen fibers have ruptured, but a few may still
be intact, giving the ligament the appearance of continuity although it is
unable to support any loads.

Loading of a joint which is unstable due to ligament or joint capsule
rupture produces abnormally high stresses on the joint cartilage. This
abnormal loading of the cartilage in the knee has been correlated with

osteoarthritis in humans and in animals (Marshall and Olsson, 1971; Alm et al., 1974).

Behavior of Bone-Ligament-Bone Complex Under Various Loading Conditions

The effects of the speed and duration of loading on the bone-ligament-bone complex are of great clinical importance both in evaluation of joint injuries and in treatment of various disorders.

Effect of Constant Loading

When a joint is subjected to constant low loading over an extended period of time, slow deformation of the soft tissues, or creep, takes place (Fig. 3–7A). This creep phenomenon (or creeping) is greatest during the first 6 to 8 hours of loading but may continue at a low rate for months. Most creeping occurs in the primary region of the load-deformation curve. The application of a constant low load to the soft tissues over a long period of time, which takes advantage of the creep phenomenon, is a useful treatment for several types of deformities. One example is manipulation of a clubfoot in a child, where the affected foot or feet are subjected to constant loads by means of a plaster cast. Another example is the treatment of idiopathic scoliosis with a plaster cast or brace, whereby constant loads are applied to elongate the soft tissues.

Correspondingly, when the soft tissues are deformed to a constant length, load relaxation takes place; that is, the load decreases with time. The greatest load relaxation takes place in the first 6 to 8 hours of loading. This effect then diminishes but may continue at a low rate for months (Fig. 3–7B).

Effect of Loading Rate

As with bone, the ligament in isolation stores more energy, requires more force to rupture, and sustains an increased elongation as the speed of loading (deformation rate) is increased (Kennedy et al., 1976). A more complex mechanical behavior is displayed when the entire bone-ligament-bone complex is tested in tension to failure.

Different parts of the bone-ligament-bone complex have been found to be strongest at different loading rates (Noyes and Grood, 1976). Anterior cruciate ligaments from knee specimens taken from 30 primates were tested in tension to failure at a slow and fast loading rate. At the slow loading rate (60 seconds), much slower than an in vivo injury mechanism, the bony insertion of the ligament was the weakest component, and a tibial spine avulsion was produced. At the fast loading rate (0.6 seconds), which simulated an in vivo injury mechanism, the ligament was the weakest component in two-thirds of the specimens tested. At the slower rate the load

Biomechanics of Collagenous Tissues

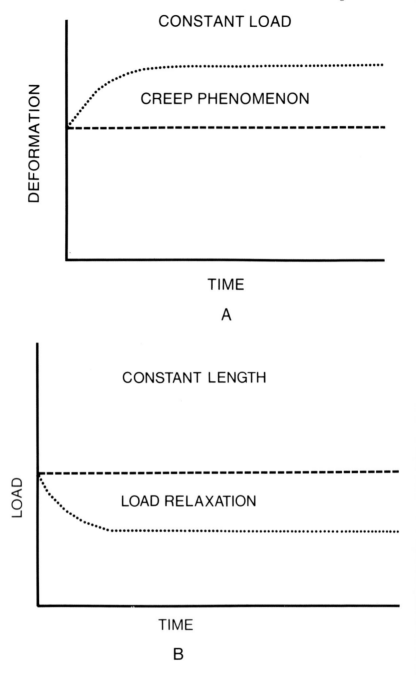

Fig. 3–7. A. The creep phenomenon takes place when a joint is subjected to a constant low load over an extended time period. Creeping of the soft tissues is greatest during the first 6 to 8 hours of loading but may continue at a low rate for months.
   B. A corresponding load relaxation of the soft tissues takes place with time.

to failure decreased by 20%, and 30% less energy was stored, but the stiffness of the bone-ligament-bone complex was almost the same. These results suggest that with an increase in loading rate, the strength of the bone increases more than the strength of the ligament.

Ligament Remodeling

Like bone, a normal ligament appears to remodel in response to the mechanical demands placed upon it. This means that a ligament becomes stronger and stiffer when subjected to increased stress and weaker and less stiff when the stress is decreased (Noyes, 1977; Tipton et al.,1970).

One situation wherein the stress on the ligament is decreased is that during immobilization of a joint. The effect of immobilization on the bone-ligament-bone complex was shown experimentally in primates who were immobilized in body casts for 8 weeks (Noyes, 1977). When tested in tension to failure, the anterior cruciate ligaments from these animals showed a 40% decrease in maximum load to failure and a significant decrease in energy storage compared with ligaments from a control group of animals (Fig. 3–8A). The immobilized ligaments were also significantly less stiff and showed increased elongation (Fig. 3–8B).

Noyes (1977) also studied the effects of a reconditioning program on the strength and stiffness of the anterior cruciate ligaments of primates immobilized for 8 weeks. After 5 months of reconditioning, the ligaments had only partially recovered their strength (in terms of maximum load to failure and energy stored) and stiffness. The strength of these ligaments was 20% less than that of the ligaments from a control group. After 12 months of reconditioning, however, the ligaments had a strength and stiffness comparable to those of control group ligaments (Fig. 3–8A).

An additional group of immobilized primates performed isotonic exercises against one-third of body weight 600 to 800 times per day during the 8-week immobilization period. This exercise regimen produced no increase in the strength and stiffness of the anterior cruciate ligaments compared with the values for the ligaments of a control group of immobilized primates.

The results of these experiments suggest that after partial or complete immobilization of a joint, a long period of time (up to 1 year) may be required before the ligament regains its normal strength and stiffness. The findings also suggest that when a joint is immobilized, an isotonic exercise program cannot simulate normal physiologic loading, and therefore cannot prevent the ligament from decreasing in strength. They do not discount, however, the many beneficial effects of an isotonic exercise program during joint immobilization which are clinically well established.

Hypertrophy of a ligament may take place in response to the increased mechanical stress imposed by strenuous exercise. Tipton et al. (1970) compared the strength and stiffness of medial collateral ligaments from dogs which were strenuously exercised for 6 weeks with the values for ligaments

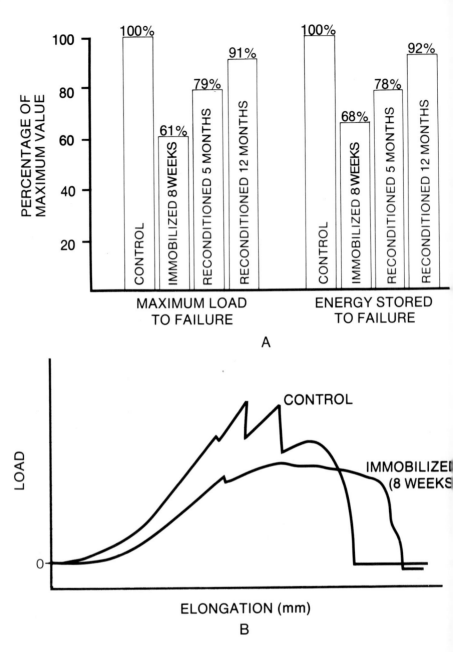

Fig. 3–8. A. Maximum load to failure and energy stored to failure for anterior cruciate ligaments tested in tension to failure. Values are shown as a percentage of control values for three groups of experimental animals:

1. Primates immobilized in body casts for 8 weeks
2. Primates immobilized for 8 weeks and given a reconditioning program for 5 months
3. Primates immobilized for 8 weeks and given a reconditioning program for 12 months
(Adapted from Noyes, 1977.)

B. Compared with controls, ligaments immobilized for eight weeks were significantly less stiff (as indicated by the slope of the curve) and showed increased elongation. (Adapted from Noyes, 1977.)

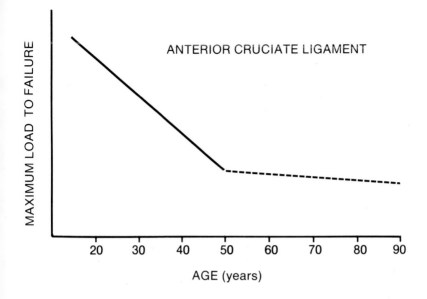

Fig. 3–9. Influence of age on maximum load to tensile failure for anterior cruciate ligaments from autopsy specimens. (Adapted from Noyes and Grood, 1976.)

from a control group of animals. The ligaments of the exercised dogs were stronger and stiffer than those of the control dogs, and the diameter of the collagen fiber bundles of these ligaments was larger.

Aging appears to produce changes in ligaments similar to those resulting from immobilization. A significant reduction in strength and stiffness of ligaments occurs with advanced age. This alteration in the mechanical properties of the ligaments may result from many factors, including degenerative processes, disuse effects related to a reduced activity level, or superimposed disease states. Studies of bone-ligament-bone specimens from older autopsy subjects showed a two- to threefold reduction in maximum load to failure, energy storage capacity, and stiffness in the anterior cruciate ligament compared with ligament specimens from young autopsy subjects (Noyes and Grood, 1976) (Fig. 3–9).

TENDONS

The function of the tendon is to attach muscle to bone or fascia, and to transmit tensile loads from muscle to bone or from muscle to fascia, thereby producing joint motion.

Two types of tendon arrangements can be identified: tendons with sheaths and those without sheaths. In certain locations where the tendons are subjected to particularly high friction forces (for example, in the dorsal parts of the palm, in the fingers, and at the level of the wrist joint), they are

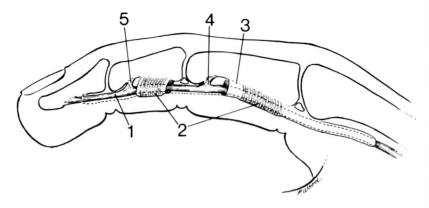

Fig. 3–10. Schematic drawing of the flexor digitorum profundus tendon (1) of the human hand showing the location of the most important pulleys (2), which are heavy fibrous condensations of the tendon sheath. The sheath (3) has been omitted in places (dotted lines) to show the tendon, the vinculum longus (4) which originates in the vinculum brevis of the superficialis tendon (not shown), and the vinculum breve (5). The primary function of the vincula is to carry a blood supply to the tendon within the tendon sheath. (Courtesy of Mary Ann Riederer-Henderson, Ph.D.)

surrounded by a sheath. This sheath is composed of a fibrous layer associated with a parietal synovial layer (Greenlee and Ross, 1967). The synovial fluid produced by the synovial cells facilitates the gliding of the tendon. In locations where the tendons are subjected to lower friction forces, they are surrounded by a peritenon, that is, a loose connective tissue.

In the finger the tendon must work or bend around corners; therefore, pulleys or annular ligaments are found in certain places as part of the sheath. Two of the most important pulleys (Doyle and Blythe, 1975) surrounding the flexor profundus tendon of the human finger are illustrated in Figure 3–10. A pulley is a simple machine in that it changes the direction of the muscular force transmitted along the tendon. The movement of the tendon over the pulleys is similar to the gliding movement of the cable of a ski lift (Verdan, 1972).

Muscle-Tendon-Bone Complex

The behavior of a tendon under loading is almost identical to that of a ligament. The same two factors determine the strength of the tendon: the size and shape of the tendon and the speed of loading. Like the ligament, the tendon cannot be considered in isolation, but must be considered as a link in the muscle-tendon-bone system.

The structure of the insertion of the tendon is similar to that of the ligament. As in the ligament, four zones can be identified (Fig. 3–11). At the end of the tendon (zone 1), the collagen fibers intermesh with fibrocartilage (zone 2). This fibrocartilage gradually becomes mineralized fibrocartilage

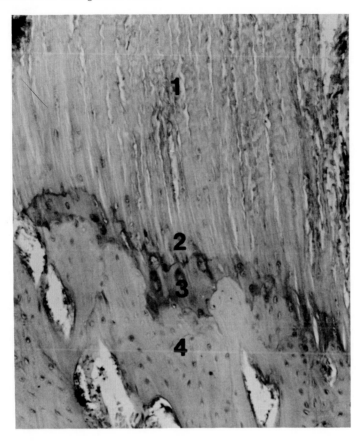

Fig. 3–11. Electron micrograph of a patellar tendon insertion from a dog, showing four zones (25,000×). Zone 1. Parallel collagen fibers; Zone 2. Unmineralized fibrocartilage; Zone 3. Mineralized fibrocartilage; Zone 4. Cortical bone. The ligament-bone junction (not pictured) has a similar appearance. (Reprinted with permission from Cooper, R. R., and Misol, S.: Tendon and ligament insertion. A light and electron microscopic study. J. Bone Joint Surg., *52A*:1–20, 1970.)

(zone 3), and then merges into cortical bone (zone 4). This change from a more tendinous material to a more bony material produces a gradual change in the mechanical properties, which results in a decreased stress concentration effect at the insertion of the tendon into the stiffer bone (Cooper and Misol, 1970).

Two main factors influence the amount of stress imposed on a tendon during activity:

1. The amount of contraction of the muscle to which the tendon is attached

2. The size of the tendon in relation to the size of its muscle

The amount of stress on a tendon increases as its muscle contracts. When the muscle is maximally contracted, the tensile stress on the tendon is high. Tensile stress on the tendon can be further increased if the muscle is rapidly extended. For example, rapid dorsiflexion of the ankle, which does not allow for reflex relaxation of the gastrocnemius and soleus muscles, increases the tension on the Achilles tendon. The load imposed on the tendon under these circumstances may exceed the yield point, causing Achilles tendon rupture.

The amount of contraction produced by a muscle depends upon its physiological cross-sectional area. The larger the cross-sectional area of the muscle is, the higher the magnitude of the force produced by the contraction, and thus the greater the tensile loads transmitted through the tendon. Similarly, the larger the cross-sectional area of the tendon is, the greater the loads it can bear. Although the maximum stress to failure for a muscle has been difficult to compute accurately, such measurements have shown that the tensile strength of a healthy tendon may be more than twice the strength of its muscle (Elliott, 1967). This finding is supported clinically by the fact that muscle ruptures are more common than ruptures through a tendon.

Large muscles usually have tendons with large cross-sectional areas. Examples are the quadriceps muscle with its patellar tendon, and the triceps surae muscle with its Achilles tendon. Some small muscles, however, have tendons with large cross-sectional areas, for example, the plantaris muscle, a tiny muscle with a large tendon.

Few studies of in vivo loading of tendons have been performed. In one of these studies, Kear and Smith (1975), using the strain gauge method, measured the maximal strain in the lateral digital extensor tendon in sheep. The strain reached 2.6% while the sheep were trotting rapidly and decreased when the trotting speed decreased. This maximal strain occurred for only 0.1 second during each stride. The maximal load imposed on the entire tendon was approximately 45 newtons. These results suggest that during normal activity a tendon in vivo is subjected to less than one-fourth of its ultimate stress. The findings of this study correlate well with those of other authors (Abrahams, 1967; Elliott, 1967).

Flexor Tendon Healing After Surgical Repair

The ability to flex the fingers of the hand is of great importance for the performance of daily activities. Thus, injuries to the sheathed flexor tendons usually require surgical repair for restoration of normal function. A main objective of this surgical repair is to restore union of the tendon ends and the normal gliding function of the tendon (Peacock, 1964).

The tendon is weakest 3 to 14 days after surgical repair. During this period, the collagen fibers become soft in the area of repair, and their holding power and ability to resist the shearing effect of the suture are decreased (Mason and Allen, 1941).

In tensile studies on dog tendons (Urbaniak et al., 1975), the strength of the tendons was found to be nearly normal immediately after repair but decreased rapidly over the next few days, reaching a minimum on the fifth day. From the sixth day the tendons gradually regained their strength, reaching nearly normal strength by the twentieth day.

Clinical experience has shown that a period of about 3 weeks of immobilization is necessary to prevent rerupturing of a surgically repaired tendon. An immobilization period of over 3 weeks often results in adhesions between the tendon and its sheath or surrounding soft tissues. Clinically, a prolonged period of immobilization is frequently found to result in joint stiffness. Both joint stiffness and the presence of adhesions can hinder the gliding function of the tendon and can result in decreased mobility of the joint. A human tendon probably does not regain its normal strength until 40 to 50 weeks after surgical repair.

Although the tendon is a passive structure that produces no active motion, it is constantly subjected to loads because of the continual contraction of its muscle. Thus, some loads may be imposed on the tendon even when the joint is immobilized.

Because immobilization does not completely eliminate the loads on the tendon, the method of producing an anastomosis surgically is of great importance. Common methods of producing anastomosis fall into four categories (Fig. 3–12). In the first category (interrupted sutures) (Fig. 3–12A), several sutures are parallel to the collagen fibers of the tendon, and thus any tensile load is applied directly to the opposing tendon ends. In the second category (Kessler suture) (Fig. 3–12B), the suture is transverse and parallel in relation to the collagen fibers of the tendon, and thus a tensile load applied to the tendon produces mostly tensile loads but also some compressive loads on the tendon ends. In the third category (Bunnell suture) (Fig. 3–12C), the suture is transverse and oblique in relation to the collagen fibers, and thus a tensile load applied to the tendon produces both compressive and tensile loads on the tendon ends. In the fourth category (fish mouth or end-weave method) (Fig. 3–12D), the suture is perpendicular to the collagen fibers, and a tensile load applied to the tendon produces shearing loads at the two anastomosis sites and compressive loads between those sites.

The end-weave method of producing anastomosis provides the greatest strength to the tendon immediately after repair (Urbaniak et al., 1975). This suture's major disadvantage is the bulkiness of the repair site. Thus, to permit gliding of the tendon, the anastomosis must be placed where the surrounding tissues are not restricted. The anastomosis produced by the Bunnell and Kessler methods immediately after repair is only 60 percent as strong as that produced by the end-weave method. The anastomosis produced by the interrupted suture method is only 25 percent as strong as that produced by the end-weave method.

In the area of hand rehabilitation, different types of exercises have been found to produce varying loads on the flexor tendons. In patients undergoing surgery for relief of median nerve compression, the tensile load on the digital

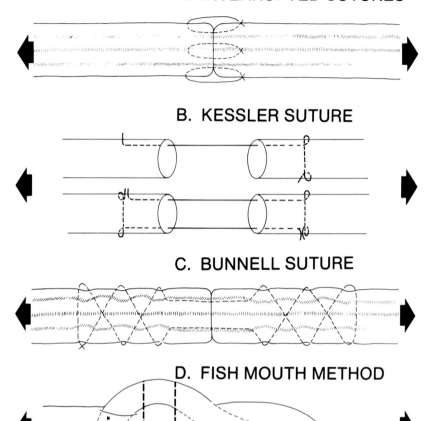

A. INTERRUPTED SUTURES

B. KESSLER SUTURE

C. BUNNELL SUTURE

D. FISH MOUTH METHOD

Fig. 3–12.   Four common methods of producing anastomosis.
   A. Interrupted sutures. The sutures are parallel to the collagen fibers.
   B. Kessler suture. The suture is transverse and parallel in relation to the collagen fibers.
   C. Bunnell suture. The suture is transverse and oblique in relation to the collagen fibers.
   D. Fish mouth method. The suture is perpendicular to the collagen fibers.

flexor tendons was measured under four exercise conditions: rest, passive flexion-extension, active flexion against slight resistance, and active flexion against moderate resistance (Urbaniak et al., 1975). Compared with rest, the loads on the tendon were twice as great with passive extension-flexion, four times as great with active flexion against slight resistance, and eight times as great with active flexion against moderate resistance. Urbaniak et al. then compared these values with the tensile loads to failure measured in dog flexor tendons 5 days after repair with various suture methods. They

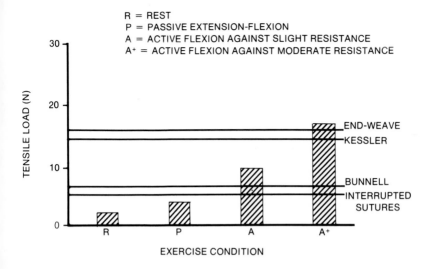

Fig. 3–13. The tensile loads on digital flexor tendons during four exercise conditions in patients undergoing hand surgery are compared with the ultimate strength of dog flexor tendons 5 days after repair with four suture methods. Only the end-weave and Kessler methods appear to provide enough strength to the tendons to allow them to withstand active motion. (Adapted from Urbaniak et al., 1975.)

concluded that at the weakest point during the repair process only the end-weave and Kessler methods of producing anastomosis would provide sufficient strength to the flexor tendons to allow them to withstand active motion (Fig. 3–13).

Tendon Remodeling

The few studies available on tendon remodeling indicate that, like the ligament, the tendon remodels in response to the amount of stress placed upon it. Tendon has been found to hypertrophy in response to increased stress (Viidik, 1967; Woo et al., 1979). In comparing the tendons of control swine with those from swine exercised strenuously for 1 year, Woo et al. found a significant increase in the strength and stiffness of the exercised tendons.

SUMMARY

1. The collagenous tissues surrounding the skeletal system are composed primarily of collagen and elastic fibers.
2. The behavior of the collagenous tissues under loading is influenced by the properties of the collagen and elastic fibers, the proportion between these fibers, and the structural orientation of the fibers.

3. Most collagenous tissues undergo deformation before failure.
4. When the collagenous tissues are subjected to constant low loading over an extended period of time, slow deformation of the tissues, or creep, takes place.
5. Like bone, the collagenous tissues increase in strength and stiffness with an increased speed of loading.
6. Collagenous tissues remodel in response to the mechanical demands placed upon them.

REFERENCES

Abrahams, M.: Mechanical behaviour of tendon *in vitro*. A preliminary report. Med. Biol. Engng., *5*:433, 1967.

Alm, A., Ekström, H., Gillquist, J., and Strömberg, B.: The anterior cruciate ligament. A clinical and experimental study on tensile strength, morphology and replacement by patellar ligament. Acta Chir. Scand., Suppl. *445*:1–49, 1974.

Carton, R. W., Dainauskas, J., and Clark, J. W.: Elastic properties of single elastic fibers. J. Appl. Physiol., *17*:547, 1962.

Cooper, R. R., and Misol, S.: Tendon and ligament insertion. A light and electron microscopic study. J. Bone Joint Surg., *52A*:1, 1970.

Doyle, J. R., and Blythe, W.: The finger flexor tendon sheath and pulleys: Anatomy and reconstruction. In *American Academy of Orthopaedic Surgeons, Symposium on Tendon Surgery in the Hand*. St. Louis, C. V. Mosby Co., 1975, pp. 81–87.

Elliott, D. H.: The biomechanical properties of tendon in relation to muscular strength. Ann. Phys. Med., *9*:1, 1967.

Fielding, J. W., Burstein, A. H., and Frankel, V. H.: The nuchal ligament. Spine, *1*:3, 1976.

Greenlee, T. K., and Ross, R.: The development of the rat flexor digital tendon, a fine structure study. J. Ultrastruct. Res., *18*:354, 1967.

Grood, E.: Personal communication, 1978.

Kear, M., and Smith, R. N.: A method for recording tendon strain in sheep during locomotion. Acta Orthop. Scand., *46*:896, 1975.

Kennedy, J. C., Hawkins, R. J., Willis, R. B., and Danylchuk, K. D.: Tension studies of human knee ligaments. Yield point, ultimate failure, and disruption of the cruciate and tibial collateral ligaments. J. Bone Joint Surg., *58A*:350, 1976.

Marshall, J. L., and Olsson, S.-E.: Instability of the knee. A long-term study in dogs. J. Bone Joint Surg., *53A*:1561, 1971.

Mason, M. L., and Allen, H. S.: The rate of healing of tendons. An experimental study of tensile strength. Ann Surg., *113*:424–459, 1941.

Nachemson, A. L., and Evans, J. H.: Some mechanical properties of the third human lumbar interlaminar ligament (ligamentum flavum). J. Biomech., *1*:211, 1968.

Noyes, F. R.: Functional properties of knee ligaments and alterations induced by immobilization. Clin. Orthop., *123*:210, 1977.

Noyes, F. R., and Grood, E. S.: The strength of the anterior cruciate ligament in humans and Rhesus monkeys. Age-related and species-related changes. J. Bone Joint Surg., *58A*:1074, 1976.

Peacock, E. E.: Fundamental aspects of wound healing relating to the restoration of gliding function after tendon repair. Surg. Gynecol. Obstet., *119*:241, 1964.

Tipton, C.M., James, S. L., Mergner, W., and Tcheng, T.: Influence of exercise on strength of medial collateral ligaments of dogs. Am. J. Physiol., *218*:894–902, 1970.

Urbaniak, J. R., Cahill, J. D., and Mortenson, R. A.: Tendon suturing methods: Analysis of tensile strengths. In *American Academy of Orthopaedic Surgeons, Symposium on Tendon Surgery in the Hand*. St. Louis, C. V. Mosby Co., 1975, pp. 70–80.

Verdan, C. E.: Half a century of flexor-tendon surgery. Current status and changing philosophies. J. Bone Joint Surg., *54A*:472–491, 1972.

Viidik, A.: The effect of training on the tensile strength of isolated rabbit tendons. Scand. J. Plast. Reconstr. Surg., *1*:141, 1967.

Woo, S. L-Y., Ritter, M. A., Sanders, T. M., Gomez, M. A., Garfin, S. R., and Akeson, W. H.: Long term exercise effects on the biomechanical and structural properties of swine tendons. Transactions of the 25th Annual Meeting, Orthopaedic Research Society, *4*:23, 1979.

SUGGESTED READING

Collagenous Tissue

Carton, R. W., Dainauskas, J., and Clark, J. W.: Elastic properties of single elastic fibers. J. Appl. Physiol., *17*:547–551, 1962.
Kivirikko, K. I., and Risteli, L.: Biosynthesis of collagen and its alterations in pathological states. Med. Biol., *54*:159–186, 1976.
Yamada, H.: *Strength of Biological Materials*. Edited by F. G. Evans. Baltimore, Williams & Wilkins, 1970.

Ligament

Alm, A., Ekström, H., Gillquist, J., and Strömberg, B.: The anterior cruciate ligament. A clinical and experimental study on tensile strength, morphology and replacement by patellar ligament. Acta Chir. Scand., Suppl. *445*:1–49, 1974.
Cooper, R. R., and Misol, S.: Tendon and ligament insertion. A light and electron microscopic study. J. Bone Joint Surg., *52A*:1–20, 1970.
Fielding, J. W., Burstein, A. H., and Frankel, V. H.: The nuchal ligament. Spine, *1*:3–14, 1976.
Kennedy, J. C., Hawkins, R. J., Willis, R. B., and Danylchuk, K. D.: Tension studies of human knee ligaments. Yield point, ultimate failure, and disruption of the cruciate and tibial collateral ligaments. J. Bone Joint Surg., *58A*:350–355, 1976.
Marshall, J. L., and Olsson, S.-E.: Instability of the knee. A long-term study in dogs. J. Bone Joint Surg., *53A*:1561–1570, 1971.
Nachemson, A. L., and Evans, J. H.: Some mechanical properties of the third human lumbar interlaminar ligament (ligamentum flavum). J. Biomech., *1*:211–220, 1968.
Noyes, F. R.: Functional properties of knee ligaments and alterations induced by immobilization. Clin. Orthop., *123*:210–242, 1977.
Noyes, F. R., and Grood, E. S.: The strength of the anterior cruciate ligament in humans and Rhesus monkeys. Age-related and species-related changes. J. Bone Joint Surg., *58A*:1074–1082, 1976.
Noyes, F. R., Torvik, P. J., Hyde, W. B., and DeLucas, J. L.: Biomechanics of ligament failure. J. Bone Joint Surg., *56A*:1406–1418, 1974.
Tipton, C. M., James, S. L., Mergner, W., and Tcheng, T.: Influence of exercise on strength of medial collateral ligaments of dogs. Am. J. Physiol., *218*:894–902, 1970.

Tendon

Abrahams, M.: Mechanical behaviour of tendon *in vitro*. A preliminary report. Med. Biol. Engng., *5*:433–443, 1967.
Blanton, P. L., and Biggs, N. L.: Ultimate tensile strengths of fetal and adult human tendons. J. Biomech., *3*:181–189, 1970.
Cooper, R. R., and Misol, S.: Tendon and ligament insertion. A light and electron microscopic study. J. Bone Joint Surg., *52A*:1–20, 1970.
Doyle, J. R., and Blythe, W.: The finger flexor tendon sheath and pulleys: Anatomy and reconstruction. In *American Academy of Orthopaedic Surgeons, Symposium on Tendon Surgery in the Hand*. St. Louis, C. V. Mosby Co., 1975, pp. 81–87.
Elliott, D. H.: The biomechanical properties of tendon in relation to muscular strength. Ann. Phys. Med., *9*:1–7, 1967.
Greenlee, T. K., and Ross, R.: The development of the rat flexor digital tendon, a fine structure study. J. Ultrastruct. Res., *18*:354–376, 1967.

Greenlee, T. K., Jr., Beckham, C., and Pike, D.: A fine structural study of the development of the chick flexor digital tendon: A model for synovial sheathed tendon healing. Am. J. Anat., *143*:303–313, 1975.
Kear, M., and Smith, R. N.: A method for recording tendon strain in sheep during locomotion. Acta Orthop. Scand., *46*:896–905, 1975.

Tendon Healing

Ketchum, L. D.: Primary tendon healing: A review. J. Hand Surg., *2*:428–435, 1977.
Lundborg, G., and Rank, F.: Experimental intrinsic healing of flexor tendons based upon synovial fluid nutrition. J. Hand Surg., *3*:21–31, 1978.
Mason, M. L., and Allen, H. S.: The rate of healing of tendons. An experimental study of tensile strength. Ann. Surg., *113*:424–459, 1941.
Matthews, P., and Richards, H.: The repair potential of digital flexor tendons. J. Bone Joint Surg., *56B*:618–625, 1974.
Peacock, E. E.: Fundamental aspects of wound healing relating to the restoration of gliding function after tendon repair. Surg. Gynecol. Obstet., *119*:241–250, 1964.
Urbaniak, J. R., Cahill, J. D., and Mortenson, R. A.: Tendon suturing methods: Analysis of tensile strengths. In *American Academy of Orthopaedic Surgeons, Symposium on Tendon Surgery in the Hand*. St. Louis, C. V. Mosby Co., 1975, pp. 70–80.
Verdan, C. E.: Half a century of flexor-tendon surgery. Current status and changing philosophies. J. Bone Joint Surg., *54A*:472–491, 1972.

Remodeling of Collagenous Tissue

Madden, J. W., and Peacock, E. E., Jr.: Studies on the biology of collagen during wound healing. III. Dynamic metabolism of scar collagen and remodelling of dermal wounds. Ann. Surg., *174*:511–520, 1971.
Noyes, F. R.: Functional properties of knee ligaments and alterations induced by immobilization. Clin. Orthop., *123*:210–242, 1977.
Noyes, F. R., and Grood, E. S.: The strength of the anterior cruciate ligament in humans and Rhesus monkeys. Age-related and species-related changes. J. Bone Joint Surg., *58A*:1074–1082, 1976.
Tipton, C. M., James, S. L., Mergner, W., and Tcheng, T.: Influence of exercise on strength of medial collateral ligaments of dogs. Am. J. Physiol., *218*:894–902, 1970.
Viidik, A.: The effect of training on the tensile strength of isolated rabbit tendons. Scand. J. Plast. Reconstr. Surg., *1*:141–147, 1967.
Woo, S. L-Y., Matthews, J. V., Akeson, W. H., Amiel, D., and Convery, F. R.: Connective tissue response to immobility. Arthritis Rheum., *18*:257–264, 1975.
Woo, S. L-Y., Ritter, M. A., Sanders, T. M., Gomez, M. A., Garfin, S. R., and Akeson, W. H.: Long term exercise effects on the biomechanical and structural properties of swine tendons. Transactions of the 25th Annual Meeting, Orthopaedic Research Society, *4*:23, 1979.
Zuckerman, J., and Stull, G. A.: Effects of exercise on knee ligament separation force in rats. J. Appl. Physiol., *26*:716–719, 1969.

Part Two

Biomechanics of Joints

CHAPTER 4

Biomechanics of the Knee

Margareta Nordin
and
Victor H. Frankel

The knee transmits loads, participates in motion, aids in conservation of momentum, and provides a force couple for activities involving the leg. The human knee is the largest and perhaps the most complex joint in the body. It is a two-joint structure composed of the tibiofemoral joint and the patellofemoral joint (Fig. 4–1). Both joints sustain higher forces (loads) than is commonly thought. The fact that the knee is a two-joint structure, sustains high forces, and is located between the body's two longest lever arms makes it particularly susceptible to injury.

This chapter will utilize the knee to introduce the basic terms, explain the methods, and demonstrate the calculations necessary for analyzing joint motion and the forces acting on a joint. This information will be applied to other joints of the body in subsequent chapters.

The knee is particularly well suited for demonstrating biomechanical analyses of joints because these analyses can be simplified in the knee and will still yield useful data. Although motion in the knee occurs simultaneously in three planes, the motion in one plane is so great that it accounts for most knee motion. Also, although muscle forces on the knee are produced by several muscles, a single muscle group (which varies with the activity performed) produces a force so large that it accounts for most of the muscle force acting on the knee. Thus, basic biomechanical analysis can be limited to motion in one plane and to the force produced by a single muscle group and can still give an understanding of knee motion and an estimation of the magnitude of the main forces on the knee.

To analyze motion in any joint one must utilize kinematic data. Kinematics is the branch of mechanics that deals with motion of a body without reference to force or mass. To analyze the forces acting on a joint one must use both kinematic and kinetic data. Kinetics is the branch of mechanics that deals with the motion of a body under the action of given forces.

KINEMATICS

Kinematic data are used to analyze the motion in a joint. Kinematics defines the range of motion and describes the surface joint motion in three

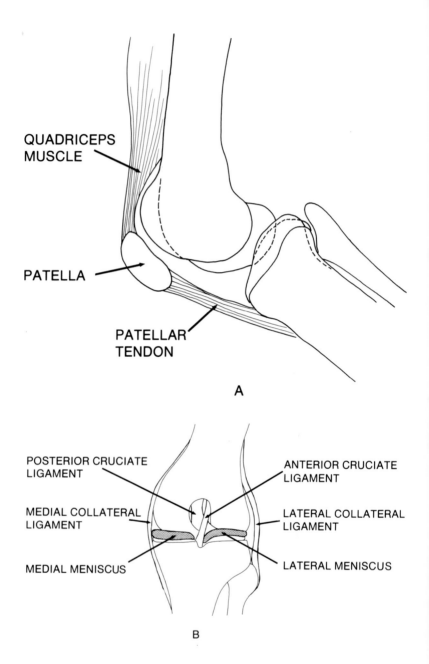

Fig. 4–1. Two-joint structure of the knee. **A.** Side view. **B.** Front view without patella.

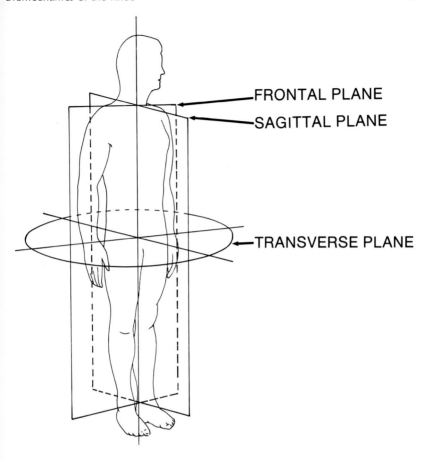

FRONTAL PLANE

SAGITTAL PLANE

TRANSVERSE PLANE

Fig. 4–2. Frontal (coronal or longitudinal), sagittal, and transverse (horizontal) planes in the human body.

planes: frontal (coronal or longitudinal), sagittal, and transverse (horizontal) (Fig. 4–2).

Range of Motion

The range of motion can be measured in any joint and in any plane. Gross measurements can be made by goniometry, but more specific measurements must be made with more precise methods such as electrogoniometry, roentgenography, or photographic techniques using skeletal pins.

In the tibiofemoral joint the range of motion is by far the greatest in the sagittal plane, where the range from full extension to full flexion of the knee is 0 to approximately 140 degrees.

In the transverse plane the range of motion in the tibiofemoral joint increases from full extension of the knee up to 90 degrees of flexion. In full extension almost no motion in this plane is possible because of the interlocking of the femoral and tibial condyles, which takes place mainly because the medial femoral condyle is longer than the lateral femoral condyle. At 90 degrees of flexion, external rotation of the knee ranges from 0 to approximately 45 degrees and internal rotation ranges from 0 to approximately 30 degrees. Beyond 90 degrees of knee flexion the range of motion in the transverse plane decreases, primarily because of the restricting function of the soft tissues.

A similar pattern is found in the frontal plane. In a fully extended knee almost no abduction or adduction is possible. When the knee is flexed up to 30 degrees, motion in this plane increases, but even at its maximum is only a few degrees in either passive abduction or passive adduction. Beyond 30 degrees of flexion, motion in this plane decreases, again because of the restricting function of the soft tissues.

The range of joint motion needed for performing various physical activities can be determined from kinematic analysis. A full range of knee motion is needed for performing the more vigorous activities of daily life in a normal manner. Any restriction of knee motion will be compensated for by increased motion in other joints.

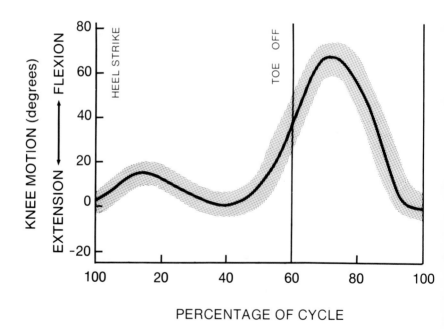

Fig. 4–3.   Range of motion of the tibiofemoral joint in the sagittal plane during level walking, one gait cycle. Shaded area indicates variation among 60 subjects (age range—20 years to 65 years). (Adapted from Murray et al., 1964.)

The range of motion of the tibiofemoral joint in the sagittal plane during level walking was measured with an electrogoniometer by Murray et al. (1964). During the entire gait cycle the knee was never fully extended. Nearly full extension (5 degrees of flexion) was noted both at the beginning of the stance phase at heel strike, and at the end of the stance phase just before toe off (Fig. 4–3). Maximum flexion (75 degrees) was observed during the middle of the swing phase.

The range of motion of the tibiofemoral joint in the transverse plane during walking was measured by several investigators. Using a photographic technique with skeletal pins through the femur and tibia, Levens et al. (1948) found that the total rotation of the tibia with respect to the femur ranged from 4.1 degrees to 13.3 degrees in 12 subjects, with a mean of 8.6 degrees. A slightly higher amount of rotation was found by Kettelkamp et al. (1970), who used an electrogoniometer to measure rotation in 22 subjects. In both studies external rotation occurred during knee extension in the stance phase and reached a peak value at the end of the swing phase just before heel strike. Internal rotation occurred during flexion in the swing phase.

Motion in the frontal plane during walking was also measured with an electrogoniometer in 22 subjects (Kettelkamp et al., 1970). In almost all subjects maximum abduction of the tibia was observed during extension at heel strike and at the beginning of the stance phase, and maximal adduction was found as the knee was flexed during the swing phase. The total amount of abduction and adduction averaged 11 degrees.

The range of motion of the tibiofemoral joint in the sagittal plane was measured during several common activities (Kettelkamp et al., 1970; Laubenthal et al., 1972) (Table 4–1). Maximal knee flexion occurred during

TABLE 4–1

RANGE OF TIBIOFEMORAL JOINT MOTION IN THE SAGITTAL PLANE DURING COMMON ACTIVITIES

Activity	Range of Motion from Knee Extension to Knee Flexion (degrees)
Walking	0– 67*
Climbing stairs	0– 83†
Descending stairs	0– 90
Sitting down	0– 93
Tying a shoe	0–106
Lifting an object	0–117

* Data from Kettelkamp et al., 1970. Mean for 22 subjects. A slight difference was found between right and left knees (mean for right knee 68.1 degrees; mean for left knee 66.7 degrees).
† These and subsequent data from Laubenthal et al., 1972. Mean for 30 subjects.

TABLE 4–2

AMOUNT OF KNEE FLEXION DURING
STANCE PHASE OF WALKING AND RUNNING

Activity	*Range in Amount of Knee Flexion during Stance Phase (degrees)*
Walking	
Slow	0– 6
Free	6–12
Fast	12–18
Running	18–30

SOURCE: Data from Perry et al., 1977. Range for seven subjects.

lifting. The values obtained for these activities indicate that full extension and at least 117 degrees of flexion are necessary for carrying out the activities of daily life in a normal manner. In one of these studies (Kettelkamp et al., 1970) a significant relationship was noted between the length of the lower leg and the range of knee motion. The longer the leg was, the greater the range of motion.

Perry et al. (1977) noted that an increase in the speed of motion required an increased range of motion in the tibiofemoral joint. From walking slowly to running, progressively greater knee flexion was needed during the stance phase (Table 4–2).

Surface Joint Motion

Surface joint motion, the motion between the articulating surfaces of a joint, can be described for any joint in the sagittal and frontal planes, but not the transverse plane. The method used is called the instant center technique. This technique allows a description of the relative uniplanar motion of two adjacent segments of a body and the direction of displacement of the contact points between these segments. Usually these segments are called links. As one link rotates about the other, there exists at an instant in time a point that does not move, that is, a point that has zero velocity. This point constitutes an instantaneous center of motion, or instant center.

The instant center for motion of a planar joint can be obtained by the method of Reuleaux (1876). According to this method, the instant center is found by identifying the displacement of two points on a link as the link moves from one position to another. The points on the link in its original position and in its displaced position are designated on a graph, and lines are drawn connecting the two sets of points. The perpendicular bisectors of these two lines are then drawn. The intersection of the perpendicular bisectors locates the instant center.

Clinically, a pathway of the instant center for a joint can be determined by taking successive roentgenograms of the joint in different positions (usually 10 degrees apart) throughout the range of motion in one plane and applying the Reuleaux method for locating the instant center for each interval of motion.

When the instant center pathway has been determined for joint motion in one plane, the surface joint motion can be described. For each interval of motion the contact point between the joint surfaces is located on the roentgenograms used for the instant center analysis, and a line is drawn from the instant center to the contact point. A second line drawn at right angles to this line will indicate the direction of displacement of the contact points. The direction of displacement of the contact points throughout the range of motion describes the surface motion in the joint. In a normal joint the line indicating the direction of displacement is tangential to the load-bearing surface, demonstrating that the femur is sliding on the tibial condyles. If the instant center were to be found on the surface, the joint would have a rolling motion and there would be no sliding friction.

Since the instant center technique allows a description of motion in one plane only, it is not useful for describing the surface joint motion if more than 15 degrees of motion takes place in any plane other than the one being measured.

In the knee, surface joint motion occurs between the tibial and femoral condyles and between the femoral condyles and the patella. Between the tibial and femoral condyles, surface motion occurs in all three planes simultaneously, but is minimal in the transverse and frontal planes. Between the femoral condyles and the patella, surface motion occurs in two planes simultaneously: the frontal and transverse.

An example will illustrate how the instant center technique is used to describe the surface motion of the tibiofemoral joint in the sagittal plane. To determine the pathway of the instant center of this joint during flexion, a lateral roentgenogram is taken of the knee in full extension, and successive films are taken at each 10 degrees of increased flexion. Care is taken to keep the tibia parallel to the x-ray table and to disallow rotation about the femur. In the case of a patient with limited knee flexion or extension, the knee is flexed or extended only as far as the patient can tolerate.

Two points on the femur that are easily identified on all roentgenograms are selected and designated on each roentgenogram (Fig. 4–4A). The films are then compared in pairs, with the images of the tibiae superimposed on each other. Roentgenograms with marked differences in tibial alignment are not utilized. Lines are drawn between the points on the femur in the two positions, and the perpendicular bisectors of these lines are then drawn. The point at which these perpendicular bisectors intersect is the instant center of

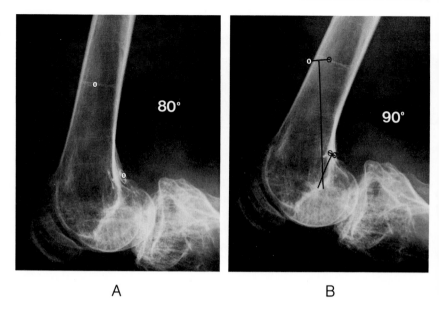

A   B

Fig. 4–4.   Locating the instant center.
  A. Two easily identifiable points on the femur are designated on a roentgenogram of a knee
  flexed 80 degrees.
  B. This roentgenogram is compared with a roentgenogram of the knee flexed 90 degrees, on
  which the same two points have been indicated. The images of the tibiae are superimposed, and lines are drawn connecting each set of points. The perpendicular bisectors of
  these two lines are then drawn. The point at which these perpendicular bisectors intersect
  locates the instant center of the tibiofemoral joint for the motion between 80 and 90
  degrees of flexion. (Courtesy of Ian Goldie, M.D.)

the tibiofemoral joint for each 10 degrees of motion (Fig. 4–4B). The instant
center pathway through the entire range of knee flexion-extension can then
be plotted. In a normal knee the instant center pathway for the tibiofemoral
joint is semicircular (Fig. 4–5).

   After the instant center pathway has been determined for the tibiofemoral
joint, the surface motion can be described. On each set of superimposed
roentgenograms the contact point between the tibiofemoral joint surfaces is
determined and a line is drawn connecting the contact point with the instant
center. A second line drawn at right angles to this line indicates the direction
of displacement of the contact points. In a normal knee this line is tangential
to the surface of the tibia for each interval of motion from full extension to
full flexion, demonstrating that the femur is sliding on the tibial condyles
(Fig. 4–6).

   Frankel et al. (1971) determined the instant center pathway and analyzed
the surface motion of the tibiofemoral joint from 90 degrees of flexion to full
extension in 25 normal knees. Tangential sliding was noted in all cases.
They also determined the instant center pathway for the tibiofemoral joint in

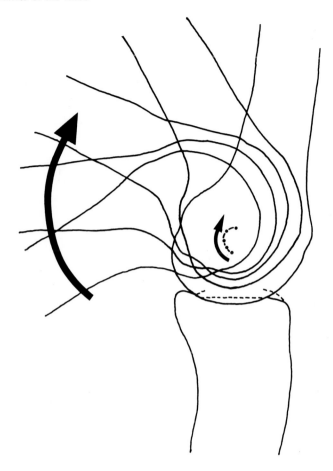

Fig. 4–5. Semicircular instant center pathway for the tibiofemoral joint in a 19-year-old man with a normal knee.

30 knees with internal derangements and found that in all cases the instant center was displaced from the normal position during some portion of the motion examined. The abnormal instant center pathway for one subject, a 35-year-old man with a "bucket handle" derangement, is shown in Figure 4–7.

If the knee is extended and flexed about a displaced instant center, the tibiofemoral joint surfaces do not slide tangentially throughout the range of motion, but become either distracted or compressed (Fig. 4–8). Such a knee is analogous to a door with a bent hinge, which no longer fits into the door jamb.

If the knee is continually forced to move about a displaced instant center, it will gradually adjust to this situation by either stretching the ligaments and

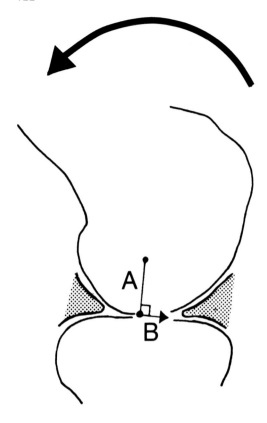

SLIDING

Fig. 4–6. In a normal knee a line drawn from the instant center of the tibiofemoral joint to the tibiofemoral contact point (line A) forms a right angle with a line tangential to the tibial surface (line B). The arrow indicates the direction of displacement of the contact points. Line B is tangential to the tibial surface, indicating that the femur slides on the tibial condyles during the measured interval of motion.

supporting structures of the joint or exerting abnormally high pressure on the articular surfaces.

Internal derangements of the tibiofemoral joint may interfere with the so-called screw-home mechanism, which is a combination of knee extension and external rotation of the tibia (Fig. 4–9). The tibiofemoral joint is not a simple hinge joint, but has spiral, or helicoid, motion. The spiral motion of the tibia about the femur during flexion and extension results from the anatomical configuration of the medial femoral condyle; in a normal knee this condyle is approximately 1.7 cm longer than the lateral femoral condyle. As the tibia slides on the femur from the fully flexed to the fully extended position, it descends and then ascends the curves of the medial femoral condyle and simultaneously rotates externally. This motion is reversed as the tibia moves back into the fully flexed position. The screw-home mechanism gives more stability to the knee in any position than would be possible if the tibiofemoral joint were a simple hinge joint.

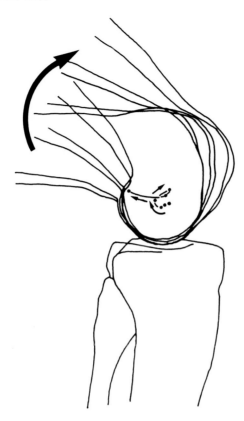

Fig. 4–7. Abnormal instant center pathway for a 35-year-old man with a "bucket handle" derangement. The instant center jumps at full extension of the knee. (Adapted from Frankel et al., 1971.)

Since the instant center technique cannot be used to analyze motion in the transverse plane, the Helfet test is used to determine if external rotation of the tibia takes place during knee extension, thereby indicating whether the screw-home mechanism is intact. This clinical test is performed with the patient sitting with 90 degrees of knee and hip flexion and the leg free-hanging. The medial and lateral borders of the patella are marked on the skin. The tibial tuberosity and the midline of the patella are then designated, and the alignment of the tibial tuberosity with the patella is checked. In a normal knee flexed 90 degrees the tibial tuberosity lines up with the medial half of the patella (Fig. 4–10A). The knee is then extended fully and the movement of the tibial tuberosity is observed. In a normal knee the tibial tuberosity moves laterally during extension and lines up with the lateral half of the patella at full extension (Fig. 4–10B). Rotatory motion in a normal knee may be as great as half the width of the patella.

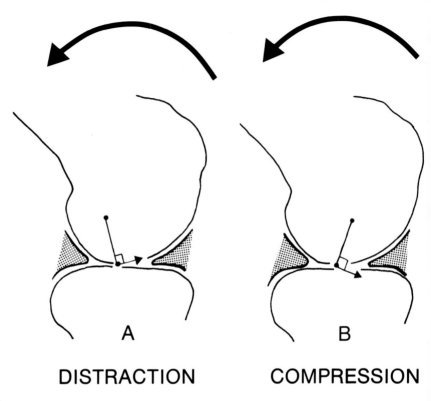

DISTRACTION   COMPRESSION

Fig. 4–8.   Surface motion in two tibiofemoral joints with displaced instant centers. In both
joints the line at right angles to the line between the instant center and the tibiofemoral
contact point indicates the direction of displacement of the contact points.
A. The small arrow indicates that with further flexion the tibiofemoral joint will be distracted.
B. The small arrow indicates that with further flexion the joint will be compressed.

In a deranged knee it may happen that no external rotation of the tibia
occurs during extension. Because of the altered surface motion, the
tibiofemoral joint will be abnormally compressed if the knee is forced into
extension, and the joint surfaces may be damaged.

Describing the surface motion of the patellofemoral joint with the instant
center technique demonstrates a sliding motion in this joint (Fig. 4–11).
From extension to full flexion of the knee the patella slides caudally
approximately 7 cm on the femoral condyles. From full extension to 90
degrees of flexion both the medial and lateral facets of the femur articulate
with the patella (Fig. 4–12). Beyond 90 degrees of flexion the patella rotates
externally, and only the medial femoral facet articulates with the patella (Fig.
4–12B). At full flexion the patella sinks into the intercondylar groove
(Goodfellow et al., 1976).

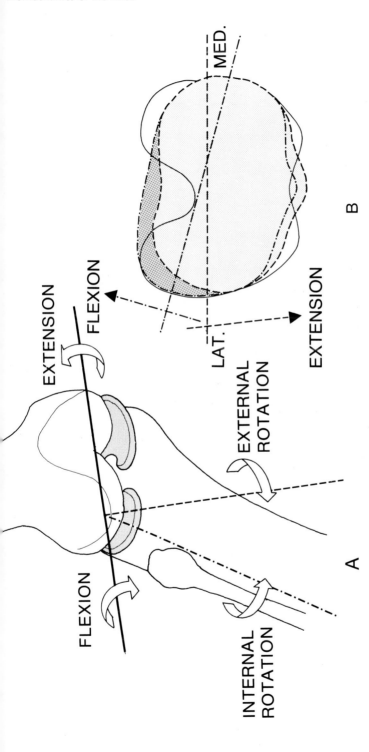

Fig. 4–9. Screw-home mechanism of the tibiofemoral joint. During knee extension the tibia rotates externally. This motion is reversed as the knee is flexed.
A. Oblique view of the femur and tibia. Shaded area indicates the tibial plateau.
B. Top view showing the position of the tibial plateau on the femoral condyles in knee flexion and extension. The lightly shaded area indicates the position of the plateau in knee flexion; the darkly shaded area indicates its position during knee extension. (Adapted from Helfet, 1974.)

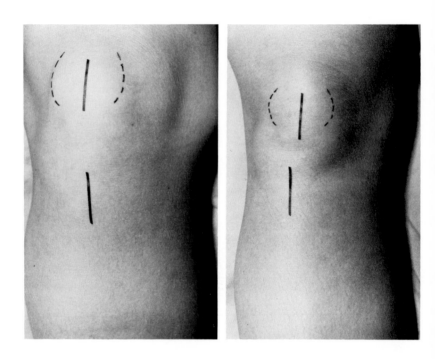

A. KNEE FLEXED 90°   B. KNEE FULLY EXTENDED

Fig. 4–10. Helfet test.

    A. In a normal knee flexed 90 degrees the tibial tuberosity lines up with the medial half of the patella.

    B. When the knee is fully extended the tibial tuberosity lines up with the lateral half of the patella.

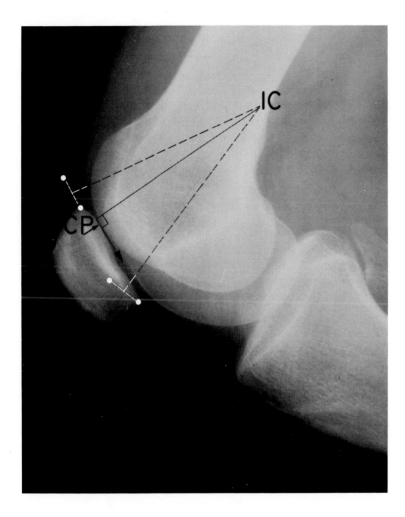

Fig. 4–11. After the instant center (IC) is determined for the patellofemoral joint for the motion from 75 to 90 degrees of knee flexion, a line is drawn from the instant center to the contact point (CP) between the patella and the femoral condyle. This line forms a right angle with a line tangential to the surface of the patella, indicating sliding.

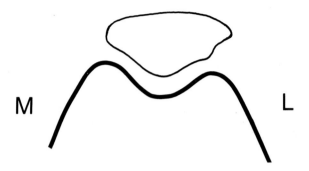

A. EXTENSION

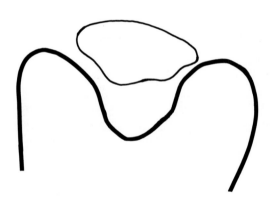

B. FLEXION > 90°

Fig. 4–12. **A.** At full extension both the medial (M) and lateral (L) femoral facets articulate with the patella.

   B. Beyond 90 degrees of flexion the patella rotates externally, and only the medial femoral facet articulates with the patella. (Adapted from Goodfellow et al., 1976.)

KINETICS

Kinetic data are used to analyze the forces acting on a joint. Kinetics involves both static and dynamic analysis. Statics is the study of forces acting on a body in equilibrium, and dynamics is the study of forces acting on a body which do not sum to zero. From kinetic analysis one can determine the magnitude of the forces on a joint produced by muscles, body weight, connective tissues, and externally applied loads in any situation, either static or dynamic, and can identify those situations that produce excessively high forces.

Statics of the Tibiofemoral Joint

Static analysis can be performed for all joints in any position and under any loading configuration. A complete static analysis involving all forces imposed on a joint is extremely complicated. For this reason a simplified technique is often used that allows the main forces acting on the joint to be analyzed. With this technique only the minimum magnitudes of these forces are obtained. The technique uses a free body diagram and limits the analysis to the three principal coplanar forces acting on the free body.

In this simplified free body technique, one portion of the body is considered as distinct from the entire body, and all forces acting on this free body are identified. A diagram is drawn of the free body in the loading situation to be analyzed. The three principal coplanar forces acting on the free body are identified and designated on the free body diagram.

These forces are designated as vectors if four characteristics are known: magnitude, sense, line of application, and point of application. (The term "direction" includes line of application and sense.) If the points of application for all three forces and the directions for two forces are known, all remaining characteristics can be obtained for an equilibrium situation. In such a situation, the three principal coplanar forces are concurrent; that is, they intersect at a common point. In other words, these forces form a closed system in which their vector sum is zero. For this reason the line of application for one force can be determined if the lines of application for the other two forces are known. Once the lines of application for all three forces are known, a triangle of forces can be constructed. From this triangle, the magnitudes of all three forces can be scaled.

An example will illustrate the application of the simplified free body technique to the knee. In this case, the technique is used to estimate the minimum magnitude of the joint reaction force acting on the tibiofemoral joint of the weight-bearing leg when the other leg is lifted during stair climbing. The lower leg is considered as a free body, distinct from the rest of the body, and a diagram of this free body in the stair-climbing situation is drawn. From all forces acting on the free body the three main coplanar forces are identified as (1) the ground reaction force (equal to body weight), (2) the tensile force through the patellar tendon exerted by the quadriceps muscle, and (3) the joint reaction force on the tibial plateau.

—The ground reaction force (W) has a known magnitude (equal to body weight*), sense, line of application, and point of application (point of contact between the foot and the ground).

—The patellar tendon force (P) has a known sense (away from the knee joint), line of application (along the patellar tendon), and point of application (point of insertion of the patellar tendon on the tibial tuberosity), but an unknown magnitude.

—The joint reaction force (J) has a known point of application on the surface of the tibia (the contact point of the joint surfaces between the tibial and femoral condyles, estimated from a roentgenogram of the joint in the proper loading configuration), but an unknown magnitude, sense, and line of application.

These three forces are designated on the free body diagram (Fig. 4–13). Because the lower limb is in equilibrium, the lines of application for all three forces will intersect at one point. Since the lines of application for two forces (forces W and P) are known, the line of application for the third force (force J) can be determined. The lines of application for forces W and P are extended until they intersect. The line of application for force J can then be drawn from its point of application on the tibial surface through the intersection point (Fig. 4–14).

Now that the line of application for force J has been determined, it is possible to construct a triangle of forces. First, a vector representing force W is drawn. Next, force P is drawn from the head of vector W. The line of application and sense of force P can be indicated, but its length cannot be determined because the magnitude is unknown. Since the lower limb is in equilibrium, however, it is known that when force J is added the triangle must close (that is, the head of force P must touch the origin of force J). The line of application of force J is then drawn from the origin of vector W. The point at which force J intersects force P is the head of vector P and the origin of vector J. The magnitudes of forces P and J can now be scaled from the drawing (Fig. 4–15). In this case the patellar tendon force (P) is 3.2 times body weight, and the joint reaction force (J) is 4.1 times body weight.

It can be seen that the main muscle force has a much greater influence on the magnitude of the joint reaction force than does the ground reaction force produced by body weight. It should be noted that in this example only the minimum magnitude of the joint reaction force has been calculated. If other muscle forces are considered, such as the force produced by contraction of the hamstrings in stabilizing the knee, the joint reaction force will increase.

Dynamics of the Tibiofemoral Joint

Although it is necessary to know how to estimate the magnitude of the forces imposed on a joint in a static situation, most of our activities are of a

---

*In this case the ground reaction force is actually equal to body weight minus the weight of the lower leg. Since the weight of the lower leg is minimal (less than one-tenth of the body weight), it can be disregarded, and the figure for total body weight can be utilized in the calculation.

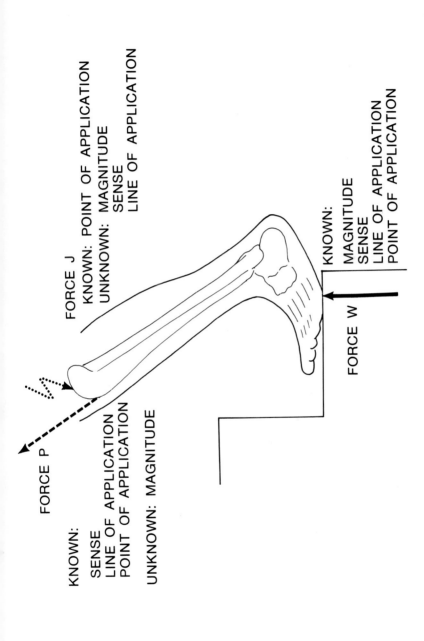

Fig. 4–13. The three main coplanar forces acting on the lower limb are designated on the free body diagram. Force W is the ground reaction force, force P is the patellar tendon force, and force J is the joint reaction force.

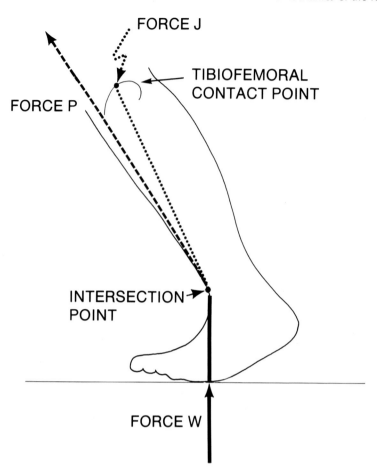

Fig. 4–14. On the free body diagram of the lower limb the lines of application for forces W and P are extended until they intersect (intersection point). The line of application for force J is then determined by connecting its point of application (tibiofemoral contact point) with the intersection point for forces W and P.

dynamic rather than a static nature. To analyze the forces acting on a joint during motion, a technique for solving dynamic problems must be used.

As in static analysis, the main forces considered in dynamic analysis are those produced by muscles, body weight, connective tissues, and externally applied loads. Friction forces, which are negligible in a normal joint, are not considered. In dynamic analysis, two factors in addition to those in static analysis must be taken into account: (1) the acceleration of the body part under consideration, and (2) the mass moment of inertia of the body part. (The mass moment of inertia is the unit used to express the amount of torque needed to accelerate a body and is dependent on the shape of the body.)

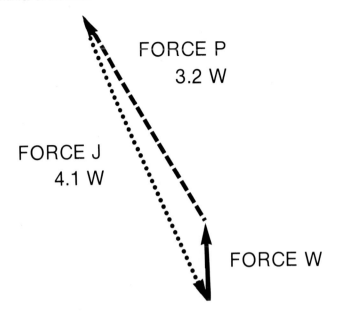

Fig. 4–15. A triangle of forces is constructed. First vector W is drawn. Next force P is drawn from the head of vector W. Then, to close the triangle, force J is drawn from the origin of vector W. The point at which forces P and J intersect defines the length of these vectors. Now that the length of all three vectors is known, the magnitudes of forces P and J can be scaled from force W, which is equal to body weight. Force P is 3.2 times body weight, and force J is 4.1 times body weight.

The steps for calculating the minimum magnitudes of the forces acting on a joint at a particular instant in time during a dynamic activity are as follows:

1. The anatomical structures involved in the production of forces are identified.
2. The angular acceleration of the moving body part is determined.
3. The mass moment of inertia of the moving body part is determined.
4. The torque acting about the joint is calculated.
5. The magnitude of the main muscle force accelerating the body part is calculated.
6. The magnitude of the joint reaction force at a particular instant in time is calculated using static analysis.

In the first step, the structures of the body involved in producing forces on the joint are identified. These are the moving body part and the main muscles in that body part that are involved in the production of the motion.

In joints of the extremities acceleration of the body part involves a change in angle. To determine the angular acceleration of the body part, the entire movement of the body part is recorded photographically. Recording can be done with a stroboscopic light and movie camera, a television scanning system, or other methods. From the films the maximal angular acceleration for a particular motion is calculated (Frankel and Burstein, 1970).

Next, the mass moment of inertia for the moving body part is determined. Anthropometric data on the body part can be used for this determination. Since calculating these data is a complicated procedure, however, tables are commonly used (Drillis et al., 1964).

The torque about the joint can now be calculated using Newton's second law of motion, which states that when motion is angular, the torque is a product of the mass moment of inertia of the body part and the angular acceleration of that part:

$$T = I\alpha,$$

where T is the torque expressed in newton meters
   I   is the mass moment of inertia expressed in newton meters times seconds squared (Nm sec²)
   α is the angular acceleration expressed in radians per second squared (r/sec²).

Not only is the torque a product of the mass moment of inertia and the angular acceleration of the body part, but it is also a product of the main muscle force accelerating the body part and the perpendicular distance of the force from the instant center of the joint (lever arm). Thus,

$$T = Fd,$$

where F is the force expressed in newtons
   d is the perpendicular distance expressed in meters.

Since T is known and d can be measured on the body part from the line of application of the force to the instant center of the joint, the equation can be solved for F. When F has been calculated, the remaining problem can be solved like a static problem using the simplified free body technique to determine the minimum magnitude of the joint reaction force acting on the joint at a certain instant in time.

An example will illustrate the use of dynamic analysis to calculate the joint reaction force on the tibiofemoral joint at a particular instant in time during a dynamic activity, that of kicking a football (Frankel and Burstein, 1970). A stroboscopic film of the knee and lower leg was taken, and the maximal angular acceleration was found to occur at the instant the foot struck the ball; the lower leg was almost vertical at this instant. From the film the maximal angular acceleration was computed to be 453 radians per second squared. From anthropometric data tables (Drillis et al., 1964) the mass moment of inertia for the lower leg was determined to be 0.35 newton meters times seconds squared. The torque about the tibiofemoral joint was calculated according to the equation torque equals mass moment of inertia times angular acceleration (T = Iα):

$$0.35 \text{ Nm sec}^2 \times 453 \text{ r/sec}^2 = 158.5 \text{ Nm}.$$

After the torque had been determined to be 158.5 newton meters and the perpendicular distance from the subject's patellar tendon to the instant center for the tibiofemoral joint had been found to be 0.05 meters, the muscle force acting on the joint through the patellar tendon was calculated using the equation torque equals force times distance (T = Fd):

$$158.5 \text{ Nm} = F \times 0.05 \text{ m}$$

$$F = \frac{158.5 \text{ Nm}}{0.05 \text{ m}}$$

$$F = 3170 \text{ N}.$$

Thus, 3,170 newtons was the maximal force exerted by the quadriceps muscle during the kicking motion.

Static analysis can now be performed to determine the minimum magnitude of the joint reaction force on the tibiofemoral joint. The main forces on this joint are identified as the patellar tendon force (P), the gravitational force of the lower leg (T), and the joint reaction force (J). The patellar tendon force (P) and the gravitational force of the lower leg (T) are known vectors. The joint reaction force (J) has an unknown magnitude, sense, and line of application. The free body technique for three coplanar forces is used to solve for J, which is found to be only slightly lower than the patellar tendon force.

As is evident from the calculations, the two main factors that influence the magnitude of the forces on a joint in dynamic situations are the acceleration of the body part and its mass moment of inertia. An increase in angular acceleration of the body part will produce a proportional increase in the torque about the joint. Although in the body the mass moment of inertia is anatomically set, it can be manipulated externally. For example, it is increased when a weight boot is applied to the foot during rehabilitative exercises of the extensor muscles of the knee. Normally a joint reaction force of approximately 50% of body weight results when the knee is extended from 90 degrees of flexion to full extension. In a person weighing 70 kg, this force is approximately 350 newtons. If a 10-kg weight boot is placed on the foot, it will exert a gravitational force of 100 newtons. This will increase the joint reaction force by 1,000 newtons, making this force almost four times larger than without the boot.

Dynamic analysis has been used to investigate the peak magnitudes of the joint reaction forces, muscle forces, and ligament forces on the tibiofemoral joint during walking. Morrison (1970) calculated the magnitude of the joint reaction force transmitted through the tibial plateau in men and women subjects during level walking. He simultaneously recorded muscle activity with electromyography to determine which muscles produced the peak magnitudes of this force on the tibial plateau during various stages of the gait cycle (Fig. 4–16).

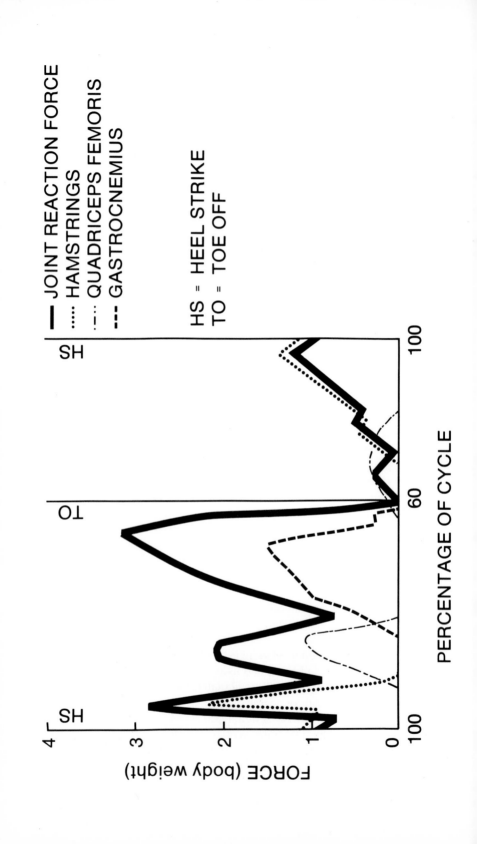

Just after heel strike the joint reaction force ranged from two to three times body weight and was associated with contraction of the hamstring muscles, which have a decelerating and stabilizing effect on the knee. During knee flexion in the beginning of the stance phase the joint reaction force was approximately two times body weight and was associated with contraction of the quadriceps muscle, which acts to prevent buckling of the knee. The peak joint reaction force occurred during the late stance phase just before toe off. This force ranged from two to four times body weight, varying among the individuals tested, and was associated with contraction of the gastrocnemius muscle. In the late swing phase contraction of the hamstring muscles resulted in a joint reaction force approximately equal to body weight. No significant difference was found between the joint reaction force magnitudes for men and women when the values were normalized by dividing them by body weight.

During the gait cycle the joint reaction force shifted from the lateral to the medial tibial plateau. In the stance phase, when the peak force occurred, it was sustained mainly by the medial plateau; in the swing phase, when the force was minimal, it was primarily sustained by the lateral plateau. The contact area of the medial tibial plateau is approximately 50% larger than that of the lateral tibial plateau (Kettelkamp and Jacobs, 1972). Also, the cartilage on this plateau is approximately three times thicker than that on the lateral plateau. The larger size and the greater thickness of the medial plateau allow it to sustain more easily the higher forces imposed upon it.

In a normal knee, joint reaction forces are sustained by the menisci as well as by the joint cartilage. The function of the menisci was investigated by Seedhom et al. (1974), who examined the distribution of stresses in knees of human autopsy subjects with and without menisci. Their results suggest that in load-bearing situations the magnitude of the stresses on the tibiofemoral joint when the menisci have been removed may be as much as three times higher than when these structures are intact.

In a normal knee stresses are distributed over a wide area of the tibial plateau. If the menisci are removed, the stresses are no longer distributed over such a wide area, but are limited to a contact area in the center of the plateau (Fig. 4–17). Thus, not only does removal of the menisci increase the magnitude of the stresses on the cartilage at the center of the tibial plateau, but it also diminishes the size and changes the location of the contact area. Over the long term the high stresses placed on this smaller contact area may be harmful to the exposed cartilage, which is usually soft and fibrillated in that area.

The forces sustained by the ligaments in the tibiofemoral joint are lower than those acting on the tibial plateau and are mainly tensile. Morrison (1970) calculated the forces on the knee ligaments during walking. The posterior cruciate ligament sustained the highest forces, about one-half body weight; peak forces occurred just after heel strike and in the later part of the stance phase.

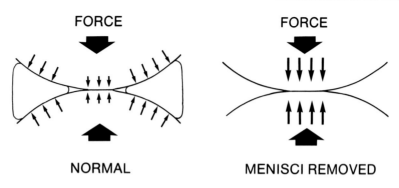

Fig. 4–17. Stress distribution in a normal knee and in a knee with the menisci removed. Removal of the menisci increases the magnitude of stresses on the cartilage of the tibial plateau and changes the size and location of the tibiofemoral contact area. With the menisci intact the contact area encompasses nearly the entire surface of the tibial plateau. With the menisci removed the contact area is limited to the center of the tibial plateau.

Function of the Patella

The patella provides two important biomechanical functions in the knee: it aids knee extension by lengthening the lever arm of the quadriceps muscle throughout the entire range of motion, and it allows a better distribution of compressive stresses on the femur by increasing the area of contact between the patellar tendon and the femur.

The contribution of the patella to the length of the quadriceps muscle lever arm changes from full flexion to full extension of the knee (Smidt, 1973; Lindahl and Movin, 1967). At full flexion, when the patella is in the intercondylar groove, it produces little anterior displacement of the quadriceps tendon, and it contributes the least to the length of the quadriceps muscle lever arm (about 10% of the total length of the lever arm). As the knee is extended, the patella rises from the intercondylar groove and produces significant anterior displacement of the tendon. The length of the quadriceps lever arm rapidly increases with extension up to 45 degrees. At this point the patella lengthens the lever arm by about 30%.

With further knee extension the length of the quadriceps lever arm decreases slightly. With this decrease in the length of its lever arm during the last 45 degrees of extension, the quadriceps muscle must exert increased force for the torque about the knee to remain the same. In an in vitro study of normal knees Lieb and Perry (1968) showed that the quadriceps muscle force required to extend the knee the last 15 degrees increased by approximately 60% (Fig. 4–18).

In a patellectomized knee the patellar tendon lies closer to the instant center of the tibiofemoral joint than in a normal knee (Fig. 4–19). Acting with a shorter lever arm, the quadriceps muscle must produce even more force than is normally required for a certain torque about the knee to be maintained during the last 45 degrees of extension. Full, active extension of

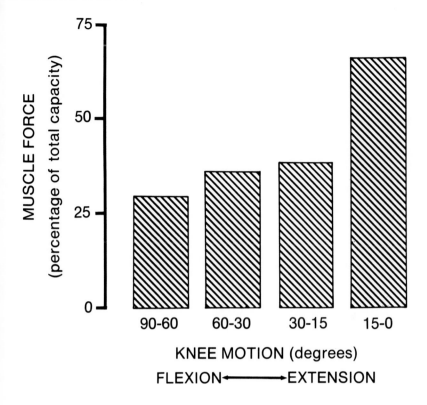

Fig. 4–18. Quadriceps muscle force required during knee motion from 90 degrees of flexion to full extension. (Adapted from Lieb and Perry, 1968.)

a patellectomized knee may require as much as 30% more quadriceps force than is normally required (Kaufer, 1971). This increased demand on the quadriceps muscle may be beyond the capacity of that muscle in some patients, particularly those with intraarticular disease or advanced age.

Kinetics of the Patellofemoral Joint

During most dynamic activities both contraction of the quadriceps muscle and body weight exert forces on the patellofemoral joint. In this situation the amount of knee flexion directly influences the magnitude of the quadriceps muscle force, which affects the magnitude of the joint reaction force. The greater the knee flexion is, the higher the magnitude of the quadriceps muscle force and, consequently, the higher the magnitude of the resultant patellofemoral joint reaction force.

During level walking, when the amount of knee flexion was relatively small, a low value was calculated for the patellofemoral joint reaction force (Reilly and Martens, 1972). The peak value for this force, which occurred

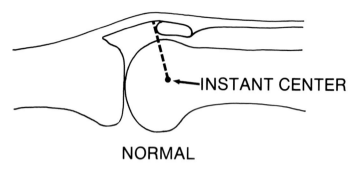

NORMAL

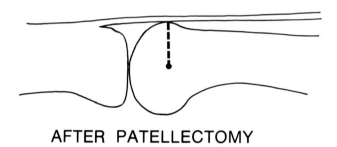

AFTER PATELLECTOMY

Fig. 4–19. Quadriceps muscle lever arm (represented by the broken line) in a normal knee and a patellectomized knee. The lever arm is the perpendicular distance between the force exerted by the quadriceps muscle through the patellar tendon and the instant center of the tibiofemoral joint for the last two degrees of extension. The patellar tendon lies closer to the instant center in the patellectomized knee. (Adapted from Kaufer, 1971.)

during the middle of the stance phase when flexion was greatest, was one-half body weight.

A much higher patellofemoral joint reaction force occurred during activities that required greater flexion. During stair climbing and descending, when knee flexion reached approximately 90 degrees, this force had a peak value of 3.3 times body weight, almost seven times the value obtained during level walking.

An even higher patellofemoral joint reaction force was produced during knee bends to 90 degrees. This force reached levels of two and one-half to three times body weight at 90 degrees of flexion (Fig. 4–20). Throughout the knee bend the patellofemoral joint reaction force remained higher than the quadriceps muscle force.

Because of the great magnitude of the patellofemoral joint reaction force during activities requiring a large amount of knee flexion, patients with

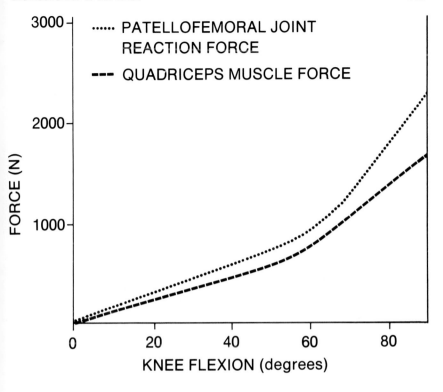

Fig. 4–20. Patellofemoral joint reaction force and quadriceps muscle force during knee bend to 90 degrees (three subjects). (Adapted from Reilly and Martens, 1972.)

patellofemoral joint derangements experience increased pain when performing these activities. An effective mechanism for decreasing the patellofemoral joint reaction force is to keep the amount of knee flexion low.

In some dynamic situations, in which body weight is minimized, a different force pattern is found. The relationship between the patellofemoral joint reaction force, the quadriceps muscle force, and the degree of knee flexion was examined for one such situation, that of knee extension against resistance produced by a weight boot with the subject in a sitting position and the lower leg free-hanging (Reilly and Martens, 1972) (Fig. 4–21). At 90 degrees of flexion the patellofemoral joint reaction force was zero. This force rapidly increased with increased knee extension, reaching a peak value of 1.4 times body weight at 36 degrees of knee flexion. With greater extension it rapidly decreased, reaching a value of approximately one-half body weight at full extension. The quadriceps muscle force was also zero at 90 degrees of flexion and increased rapidly with increased extension, reaching a peak value at full extension. If manual resistance is provided at a right angle to the long axis of the lower limb throughout knee extension, the

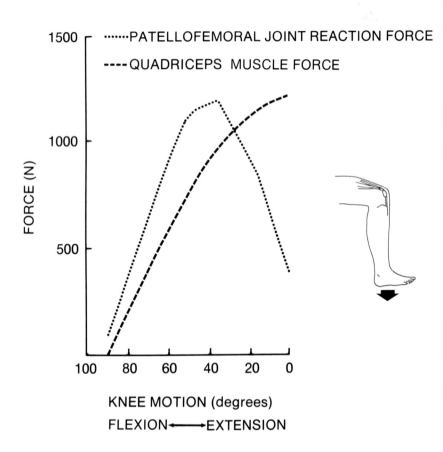

Fig. 4–21. Patellofemoral joint reaction force and quadriceps muscle force during knee extension against resistance provided by a 9-kg weight boot with the subject sitting and the lower leg free-hanging (three subjects). (Adapted from Reilly and Martens, 1972.)

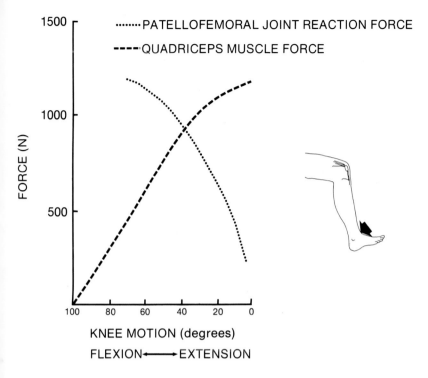

Fig. 4–22. Patellofemoral joint reaction force and quadriceps muscle force during knee extension with resistance provided at a right angle to the long axis of the lower limb throughout knee extension.

patellofemoral joint reaction force is 1.4 times body weight at 90 degrees of flexion and steadily decreases throughout extension (Fig. 4–22). The fact that the patellofemoral joint reaction force is low at full extension explains why patients with patellofemoral joint derangements are able to perform exercises against resistance with less pain if knee flexion is kept lower than 20 degrees.

SUMMARY

1. The knee is a two-joint structure composed of the tibiofemoral joint and the patellofemoral joint.
2. In the tibiofemoral joint, surface motion occurs in three planes simultaneously but is by far the greatest in the sagittal plane. In the patellofemoral joint, surface motion occurs simultaneously in two planes and is greater in the frontal than the transverse plane.
3. The surface joint motion can be described through an instant center technique. When the technique is performed on a normal knee, it shows that (a) the instant center for successive intervals of motion in

the sagittal plane follows a semicircular pathway, and (b) the direction of displacement of the tibiofemoral contact points is tangential to the surface of the tibia, indicating sliding throughout the range of motion.

4. The screw-home mechanism of the tibiofemoral joint adds stability to the joint in full extension.

5. Both the tibiofemoral and the patellofemoral joints are subjected to high forces. The magnitude of the joint reaction force on both joints can reach several times body weight. Muscle forces have the greatest influence on the magnitude of the joint reaction force.

6. Although the tibial plateaus are the main load-bearing structures in the knee, the cartilage, menisci, and ligaments also bear loads. The menisci aid in distributing the stresses imposed on the tibial plateaus.

7. The patella aids knee extension by lengthening the lever arm of the quadriceps muscle throughout the entire range of motion and allows a better distribution of compressive stresses on the femur.

REFERENCES

Drillis, R., Contini, R., and Bluestein, M.: Body segment parameters. A survey of measurement techniques. Artif. Limbs, *8*:44, 1964.

Frankel, V. H., and Burstein, A. H.: *Orthopaedic Biomechanics.* Philadelphia, Lea & Febiger, 1970.

Frankel, V. H., Burstein, A. H., and Brooks, D. B.: Biomechanics of internal derangement of the knee. Pathomechanics as determined by analysis of the instant centers of motion. J. Bone Joint Surg., *53A*:945, 1971.

Goodfellow, J., Hungerford, D. S., and Zindel, M.: Patello-femoral joint mechanics and pathology. I. Functional anatomy of the patello-femoral joint. J. Bone Joint Surg., *58B*:287, 1976.

Helfet, A. J.: Anatomy and mechanics of movement of the knee joint. In *Disorders of the Knee.* Edited by A. Helfet, Philadelphia, J. B. Lippincott, 1974, pp. 1–17.

Kaufer, H.: Mechanical function of the patella. J. Bone Joint Surg., *53A*:1551, 1971.

Kettelkamp, D. B., and Jacobs, A. W.: Tibiofemoral contact area—determination and implications. J. Bone Joint Surg., *54A*:349, 1972.

Kettelkamp, D. B., Johnson, R. J., Smidt, G. L., Chao, E. Y. S., and Walker, M.: An electrogoniometric study of knee motion in normal gait. J. Bone Joint Surg., *52A*:775, 1970.

Laubenthal, K. N., Smidt, G. L., and Kettelkamp, D. B.: A quantitative analysis of knee motion during activities of daily living. Phys. Ther., *52*:34, 1972.

Levens, A. S., Inman, V. T., and Blosser, J. A.: Transverse rotation of the segments of the lower extremity in locomotion. J. Bone Joint Surg., *30A*:859, 1948.

Lieb, F. J., and Perry, J.: Quadriceps function. An anatomical and mechanical study using amputated limbs. J. Bone Joint Surg., *50A*:1535, 1968.

Lindahl, O., and Movin, A.: The mechanics of extension of the knee-joint. Acta Orthop. Scand., *38*:226, 1967.

Morrison, J. B.: The mechanics of the knee joint in relation to normal walking. J. Biomech., *3*:51, 1970.

Murray, M. P., Drought, A. B., and Kory, R. C.: Walking patterns of normal men. J. Bone Joint Surg., *46A*:335, 1964.

Perry, J., Norwood, L., and House, K.: Knee posture and biceps and semimembranosis muscle action in running and cutting (an EMG study). Transactions of the 23rd Annual Meeting, Orthopaedic Research Society, *2*:258, 1977.

Reilly, D. T., and Martens, M.: Experimental analysis of the quadriceps muscle force and patello-femoral joint reaction force for various activities. Acta Orthop. Scand., *43*:126, 1972.

Reuleaux, F.: *The Kinematics of Machinery: Outline of a Theory of Machines.* London, Macmillan, 1876.

Seedhom, B. B., Dowson, D., and Wright, V.: The load-bearing function of the menisci: A preliminary study. In *The Knee Joint. Recent Advances in Basic Research and Clinical Aspects.* Edited by O. S. Ingwersen et al. Amsterdam, Excerpta Medica, 1974, pp. 37–42.

Smidt, G. L.: Biomechanical analysis of knee flexion and extension. J. Biomech., 6:79, 1973.

SUGGESTED READING

Tibiofemoral Joint

Brantigan, O. C., and Voshell, A. F.: The mechanics of the ligaments and menisci of the knee joint. J. Bone Joint Surg., 23A:44–66, 1941.

Brattström, H. H., Junerfält, I., and Moritz, U.: Behandlingen av sträckdefekter i leder: Kontraktur eller fraktur? Läkartidningen, 68:304–306, 1971.

Cailliet, R.: *Knee Pain and Disability.* Philadelphia, F. A. Davis Co., 1968.

Collopy, M. C., Murray, M. P., Gardner, G. M., DiUlio, R. A., and Gore, D. R.: Kinesiologic measurements of functional performance before and after geometric total knee replacement. One-year follow-up of twenty cases. Clin. Orthop., 126:196–202, 1977.

Cox, J. S., Nye, C. E., Schaefer, W. W., and Woodstein, I. J.: The degenerative effects of partial and total resection of the medial meniscus in dogs' knees. Clin. Orthop., 109:178–183, 1975.

Detenbeck, L. C.: Function of the cruciate ligaments in knee stability. J. Sports Med., 2:217–221, 1974.

Drillis, R., Contini, R., and Bluestein, M.: Body segment parameters. A survey of measurement techniques. Artif. Limbs, 8:44–66, 1964.

Ducroquet, R., Ducroquet, J., and Ducroquet, P.: *Walking and Limping. A Study of Normal and Pathological Walking.* Philadelphia, J. B. Lippincott Co., 1968.

Edholm, P., Lindahl, O., Lindholm, B., Myrnerts, R., Olsson, K.-E., and Wennberg, E.: Knee instability. An orthoradiographic study. Acta Orthop. Scand., 47:658–663, 1976.

Fairbank, T. J.: Knee joint changes after meniscectomy. J. Bone Joint Surg., 30B:664–670, 1948.

Frankel, V. H., and Burstein, A. H.: *Orthopaedic Biomechanics.* Philadelphia, Lea & Febiger, 1970.

Frankel, V. H., Burstein, A. H., and Brooks, D. B.: Biomechanics of internal derangement of the knee. Pathomechanics as determined by analysis of the instant centers of motion. J. Bone Joint Surg., 53A:945–962, 1971.

Hainaut, K.: Introduction à la biomecanique. Brussels, Presses Universitaires de Bruxelles, 1971.

Hallen, L. G., and Lindahl, O.: The "screw-home" movement in the knee-joint. Acta Orthop. Scand., 37:97–106, 1966.

Helfet, A.: *Disorders of the Knee.* Philadelphia, J. B. Lippincott Co., 1974.

Hsieh, H.-H., and Walker, P. S.: Stabilizing mechanisms of the loaded and unloaded knee joint. J. Bone Joint Surg., 58A:87–93, 1976.

Ingwersen, O. S., Van Linge, B., Van Rhens, Th. J. G., Rösingh, G. E., Veraart, B. E. E. M. J., and LaVay, D. (eds): *The Knee Joint. Recent Advances in Basic Research and Clinical Aspects.* Amsterdam, Excerpta Medica, 1974.

Johnson, R. J., Kettelkamp, D. B., Clark, W., and Leaverton, P.: Factors affecting late results after meniscectomy. J. Bone Joint Surg., 56A:719–729, 1974.

Kapandji, I. A.: *The Physiology of the Joints.* Vol. 2. Lower Limb. Edinburgh, London, and New York, Churchill Livingstone, 1970.

Kettelkamp, D. B., and Jacobs, A. W.: Tibiofemoral contact area—Determination and implications. J. Bone Joint Surg., 54A:349–356, 1972.

Kettelkamp, D. B., Johnson, R. J., Smidt, G. L., Chao, E. Y. S., and Walker, M.: An electrogoniometric study of knee motion in normal gait. J. Bone Joint Surg., 52A:775–790, 1970.

Krause, W. R., Pope, M. H., Johnson, R. J., and Wilder, D. G.: Mechanical changes in the knee after meniscectomy. J. Bone Joint Surg., 58A:599–604, 1976.

Laasonen, E. M., and Wilppula, E.: Why a meniscectomy fails. Acta Orthop. Scand., *47*:672–675, 1976.
Laubenthal, K. N., Smidt, G. L., and Kettelkamp, D. B.: A quantitative analysis of knee motion during activities of daily living. Phys. Ther., *52*:34–42, 1972.
Levens, A. S., Inman, V. T., and Blosser, J. A.: Transverse rotation of the segments of the lower extremity in locomotion. J. Bone Joint Surg., *30A*:859–872, 1948.
Lindahl, O., and Movin, A.: The mechanics of extension of the knee-joint. Acta Orthop. Scand., *38*:226–324, 1967.
Markolf, K. L., Mensch, J. S., and Amstutz, H. C.: Stiffness and laxity of the knee—The contributions of the supporting structures. J. Bone Joint Surg., *58A*:583–593, 1976.
Marshall, J. L., and Olsson, S.-E.: Instability of the knee. A long-term experimental study in dogs. J. Bone Joint Surg., *53A*:1561–1570, 1971.
McLeod, P. C., Kettelkamp, D. B., Srinivasan, V., and Henderson, O. L.: Measurements of repetitive activities of the knee. J. Biomech., *8*:369–373, 1975.
McLeod, W. D., Moschi, A., Andrews, J. R., and Hughston, J. C.: Tibial plateau topography. Am. J. Sports Med., *5*:13–18, 1977.
Moore, T. M., Meyers, M. H., and Harvey, P. J.: Collateral ligament laxity of the knee. J. Bone Joint Surg., *58A*:594–598, 1976.
Morrison, J. B.: Bioengineering analysis of force actions transmitted by the knee joint. Bio-medical Engng., *3*:164–171, 1968.
———: The mechanics of the knee joint in relation to normal walking. J. Biomech., *3*:51–61, 1970.
Murray, M.: Gait as a total pattern of movement. Including a bibliography on gait. Am. J. Phys. Med., *46*:290–298, 1967.
Murray, M. P., Drought, A. B., and Kory, R. C.: Walking patterns of normal men. J. Bone Joint Surg., *46A*:335–360, 1964.
Perry, J., Norwood, L., and House, K.: Knee posture and biceps and semimembranosis muscle action in running and cutting (an EMG study). Transactions of the 23rd Annual Meeting, Orthopaedic Research Society, *2*:258, 1977.
Pope, M. H., Crowninshield, R., Miller, R., and Johnson, R.: The static and dynamic behavior of the human knee *in vivo*. J. Biomech., *9*:449–452, 1976.
Reuleaux, F.: *The Kinematics of Machinery: Outline of a Theory of Machines*. London, Macmillan, 1876.
Roberts, E. M., Zernicke, R. F., Youm, Y., and Huang, T. C.: Kinetic parameters of kicking. In *Biomechanics IV.* Edited by R. C. Nelson and C. A. Morehouse. Baltimore, University Park Press, 1974, pp. 157–162.
Saunders, J. B. DeC. M., Inman, V. T., and Eberhardt, H. D.: The major determinants in normal and pathological gait. J. Bone Joint Surg., *35A*:543–558, 1953.
Seedhom, B. B., Dowson, D., and Wright, V.: The load-bearing function of the menisci: A preliminary study. In *The Knee Joint. Recent Advances in Basic Research and Clinical Aspects.* Edited by O. S. Ingwersen et al. Amsterdam, Excerpta Medica, 1974, pp. 37–42.
Seireg, A., and Arvikar, R. J.: The prediction of muscular load sharing and joint forces in the lower extremities during walking. J. Biomech., *8*:89–102, 1975.
Shrive, N. G., O'Connor, J. J., and Goodfellow, J. W.: Load-bearing in the knee joint. Clin. Orthop., *131*:279–287, 1978.
Smidt, G. L.: Biomechanical analysis of knee flexion and extension. J. Biomech., *6*:79–92, 1973.
Smillie, T.: *Injuries of the Knee Joint.* Edinburgh, London, and New York, Churchill Livingstone, 1970.
Stauffer, R. N., Chao, E. Y. S., and Györy, A. N.: Biomechanical gait analysis of the diseased knee joint. Clin. Orthop., *126*:246–255, 1977.
Townsend, M. A., Izak, M., and Jackson, R. W.: Total motion knee goniometry. J. Biomech., *10*:183–193, 1977.
Trent, P. S., Walker, P. S., and Wolf, B.: Ligament length patterns, strength, and rotational axes of the knee joint. Clin. Orthop., *117*:263–270, 1976.
Walker, P. S., and Erkman, M. J.: The role of the menisci in the force transmission across the knee. Clin. Orthop., *109*:184–192, 1975.
Walker, P. S., and Hajek, J. V.: The load-bearing area in the knee joint. J. Biomech., *5*:581–589, 1972.
Wang, C.-J., and Walker, P. S.: Rotatory laxity of the human knee joint. J. Bone Joint Surg., *56A*:161–170, 1974.

Wang, C.-J., Walker, P. S., and Wolf, B.: The effects of flexion and rotation on the length patterns of the ligaments of the knee. J. Biomech., 6:587–596, 1973.
Warren, C. G., Lehmann, J. F., and Kirkpatrick, G. S.: Measurement of moments in the knee-ankle orthosis of ambulating paraplegics. In *Biomechanics IV*. Edited by R. C. Nelson and C. A. Morehouse. Baltimore, University Park Press, 1974, pp. 409–414.
Williams, M., and Lissner, H.: *Biomechanics of Human Motion*. Edited by B. LeVeau, 2nd ed. Philadelphia, W. B. Saunders Co., 1977.

Patellofemoral Joint

Bandi, W.: Die retropatellaren Kniegelenk-schäden. Pathomechanik und pathologische Anatomie, Klinik und Terapie. Aktuelle Probleme in Chirurgie und Orthopädie, Band 4. Edited by M. Saegesser. Bern, Verlag Hans Huber, 1977.
Böstrom, A.: Fracture of the patella. A study of 422 patellar fractures. Acta Orthop. Scand., Suppl. *143*:1–80, 1972.
Brattström, H. H., Junerfält, I., and Moritz, U.: Behandlingen av sträckdefekter i leder: Kontraktur eller fraktur? Läkartidningen, *68*:304–306, 1971.
Cailliet, R.: *Knee Pain and Disability*. Philadelphia, F. A. Davis Co., 1968.
Drillis, R., Contini, R., and Bluestein, M.: Body segment parameters. A survey of measurement techniques. Artif. Limbs, *8*:44–66, 1964.
Frankel, V. H., and Burstein, A. H.: *Orthopaedic Biomechanics*. Philadelphia, Lea & Febiger, 1970.
Frankel, V. H., Burstein, A. H., and Brooks, D. B.: Biomechanics of internal derangement of the knee. J. Bone Joint Surg., *53A*:945–962, 1971.
Goodfellow, J., Hungerford, D. S., and Zindel, M.: Patello-femoral joint mechanics and pathology. I. Functional anatomy of the patello-femoral joint. J. Bone Joint Surg., *58B*:287–290, 1976.
Hainaut, K.: *Introduction à la biomecanique*. Brussels, Presses Universitaires de Bruxelles, 1971.
Helfet, A.: *Disorders of the Knee*. Philadelphia, J. B. Lippincott Co., 1974.
Kapandji, I. A.: *The Physiology of the Joints*. Vol. 2. Lower Limb. Edinburgh, London, and New York. Churchill Livingstone, 1970.
Kaufer, H.: Mechanical function of the patella. J. Bone Joint Surg., *53A*:1551–1560, 1971.
Reilly, D. T., and Martens, M.: Experimental analysis of the quadriceps muscle force and patello-femoral joint reaction force for various activities. Acta Orthop. Scand., *43*:126–137, 1972.
Seireg, A., and Arvikar, R. J.: The prediction of muscular load sharing and joint forces in the lower extremities during walking. J. Biomech., *8*:89–102, 1975.
Smidt, G. L.: Biomechanical analysis of knee flexion and extension. J. Biomech., 6:79–92, 1973.
Smillie, T.: *Injuries of the Knee Joint*. Edinburgh, London, and New York, Churchill Livingstone, 1970.
West, F. E.: End results of patellectomy. J. Bone Joint Surg., *44A*:1089–1108, 1962.
Williams, M., and Lissner, H.: *Biomechanics of Human Motion*. Edited by B. LeVeau, 2nd ed. Philadelphia, W. B. Saunders Co., 1977.

Quadriceps Muscle

Damholt, V., and Zdravkovic, D.: Quadriceps function following fractures of the femoral shaft. Acta Orthop. Scand., *43*:148–156, 1972.
Elftman, H.: Biomechanics of muscle. With particular application to studies of gait. J. Bone Joint Surg., *48A*:363–377, 1966.
Frankel, V. H., and Burstein, A. H.: *Orthopaedic Biomechanics*. Philadelphia, Lea & Febiger, 1970.
Haffajee, D., Moritz, U., and Svantesson, G.: Isometric knee extension strength as a function of joint angle, muscle length and motor unit activity. Acta Orthop. Scand., *43*:138–147, 1972.
Helfet, A.: *Disorders of the Knee*. Philadelphia, J. B. Lippincott Co., 1974.

Lieb, F. J., and Perry, J.: Quadriceps function. An anatomical and mechanical study using amputated limbs. J. Bone Joint Surg., *50A*:1535–1548, 1968.

———: Quadriceps function. An electromyographic study under isometric conditions. J. Bone Joint Surg., *53A*:749–758, 1971.

Reilly, D. T., and Martens, M.: Experimental analysis of the quadriceps muscle force and patello-femoral joint reaction force for various activities. Acta Orthop. Scand., *43*:126–137, 1972.

Smillie, I. S.: *Injuries of the Knee Joint.* Edinburgh, London, and New York, Churchill Livingstone, 1970.

Williams, M., and Lissner, H.: *Biomechanics of Human Motion.* Edited by B. LeVeau, 2nd ed. Philadelphia, W. B. Saunders Co., 1977.

Zernicke, R. F., Garhammer, J., and Jobe, F. W.: Human patellar-tendon rupture. J. Bone Joint Surg., *59A*:179–183, 1977.

CHAPTER 5

Biomechanics of the Hip

Margareta Nordin
and
Victor H. Frankel

Unlike the knee joint, the hip joint has intrinsic stability because of its ball-and-socket configuration. Derangements of this ball-and-socket joint can produce altered stress distributions in the joint cartilage and bone, leading to degenerative arthritis. This damage to the joint is further potentiated by the large forces borne by the joint. Although the hip joint has great inherent stability, it also has a great deal of mobility which allows normal locomotion in the performance of daily activities.

ANATOMICAL CONSIDERATIONS

The hip joint is a ball-and-socket joint composed of the acetabulum and femoral head (Fig. 5–1). The stable hip joint is constructed to allow for a large range of motion necessary for normal daily activities such as walking, sitting, and squatting. Such a joint must be precisely aligned and controlled.

The Acetabulum

The acetabulum is the concave component of the ball-and-socket configuration of the hip joint. The acetabular surface is covered with cartilage which thickens peripherally (Kempson et al., 1971). The cavity of the acetabulum faces obliquely forward, outward, and downward. A plane through the circumference of the acetabulum at its opening would intersect with the sagittal plane at an angle of 40 degrees opening posteriorly, and with a transverse plane at an angle of 60 degrees opening laterally. The acetabular cavity is deepened by the labrum.

The Femoral Head

The femoral head is the convex component of the ball-and-socket configuration of the hip joint and forms two-thirds of a sphere. The cartilage covering the femoral head is thickest on the medial-central surface and thinnest towards the periphery. The variations in the thickness of the

149

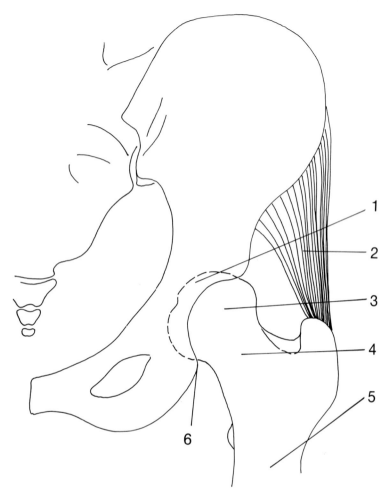

Fig. 5–1. The hip joint (front view): 1. acetabulum; 2. gluteus medius muscle; 3. femoral head; 4. femoral neck; 5. femoral shaft; 6. labrum.

cartilage result in a different strength and stiffness in different regions of the femoral head (Kempson et al., 1971). These differences in the mechanical properties from point to point on the cartilage of the femoral head may influence the transmission of stresses from the acetabulum through the femoral head to the femoral neck. Although it is not known just how stresses on the femoral head are distributed, the joint reaction force usually acts on the superior quadrant (Rydell, 1965).

The Femoral Neck

The femoral neck has two important angular relationships with the femoral shaft: the angle of inclination of the neck to the shaft in the frontal

NECK-SHAFT ANGLE

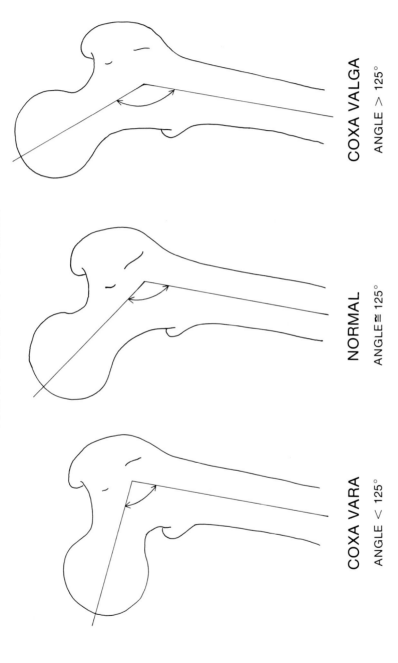

COXA VARA

ANGLE < 125°

NORMAL

ANGLE ≅ 125°

COXA VALGA

ANGLE > 125°

Fig. 5–2. The normal neck-shaft angle (angle of inclination of the femoral neck to the shaft in the frontal plane) is approximately 125 degrees. The condition wherein this angle is less than 125 degrees is termed coxa vara. If the angle is greater than 125 degrees, the condition is termed coxa valga.

plane, the neck-shaft angle, and the angle of inclination in the transverse plane, the angle of anteversion. Freedom of motion of the hip joint is facilitated by the neck-shaft angle, which places the femoral shaft away from the pelvis laterally. In most adults this angle is about 125 degrees, but it can vary from 90 to 135 degrees. If the angle is greater than 125 degrees, the condition is termed coxa valga; if the angle is less than 125 degrees, the condition is coxa vara (Fig. 5–2). Deviation of the femoral shaft into either coxa valga or vara will alter the force relationships about the hip joint.

The angle of anteversion is formed as a projection of the long axis of the femoral head and the transverse axis of the femoral condyle. In adult populations this angle averages about 12 degrees, but may vary widely. Anteversion of more than 12 degrees will cause a portion of the femoral head to be uncovered and will create a tendency toward internal rotation of the leg during gait to keep the femoral head in the acetabular cavity. An angle of less than 12 degrees will produce a tendency toward external rotation of the leg during gait. Both anteversion and retroversion are fairly common in children, but are usually outgrown.

The interior of the femoral neck is composed of cancellous bone which is divided into the medial and lateral trabecular systems (Fig. 5–3). The joint reaction force on the head of the femur parallels the trabeculae of the medial trabecular system (Frankel, 1960), indicating that this system is important for supporting the joint reaction force. It is likely that the lateral trabecular system resists the compressive force produced by contraction of the abductor muscles. The epiphyseal plates are at right angles to the trabeculae of the medial trabecular system, and are thought to be perpendicular to the joint reaction force on the femoral head (Inman, 1947). The thin shell of cortical bone around the superior femoral neck gradually thickens in the inferior region. With the aging process the neck of the femur gradually undergoes degenerative changes wherein the cortical bone is thinned and cancellated and the trabeculae gradually resorb. These degenerative changes may predispose the femoral neck to fracture.

KINEMATICS

In considering kinematics of the hip joint, it is useful to view the joint as a stable ball-and-socket configuration wherein the femoral head and acetabulum have mobility in all directions.

Range of Motion

Hip motion takes place in all three planes: sagittal, frontal, and transverse. The largest range of motion in the hip joint is found in the sagittal plane, where the range of flexion is from 0 to approximately 140 degrees and the range of extension is from 0 up to 15 degrees. In the frontal plane abduction ranges from 0 to 30 degrees, while the range of adduction is found to be

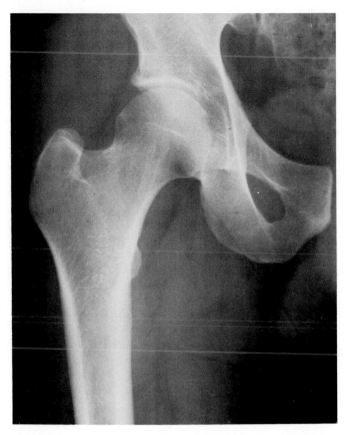

Fig. 5–3. Roentgenogram of a femoral neck showing the medial and lateral trabecular systems. The thin shell of cortical bone around the superior femoral neck gradually thickens in the inferior region.

somewhat less, from 0 to 25 degrees. In the transverse plane external rotation ranges from 0 to 90 degrees and internal rotation ranges from 0 to 70 degrees when the hip joint is flexed. Less rotation occurs when the hip joint is extended because of the restricting function of the soft tissues.

The range of motion of the hip joint in the sagittal plane during level walking was measured with an electrogoniometer by Murray (1967). The hip joint is maximally flexed during gait in late swing phase as the limb moves forward for heel strike. As the body moves forward at the beginning of stance phase, the hip joint extends. Maximum extension is reached at heel off. The joint reverses into flexion during swing phase and again reaches maximal flexion, 35 to 40 degrees, prior to heel strike. Figure 5–4 shows hip joint motion in the sagittal plane during a gait cycle and allows a comparison of this motion with that of the knee and ankle.

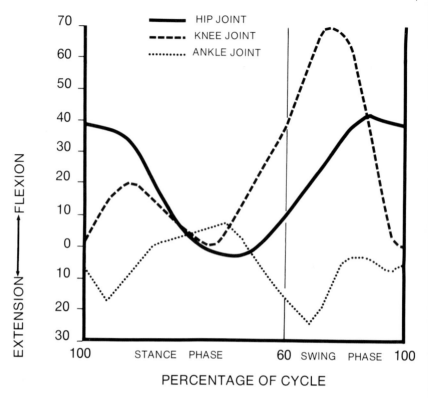

Fig. 5–4. Range of hip joint motion in the sagittal plane for 30 normal men during level walking, one gait cycle. The ranges of motion for the knee and ankle joints are shown for comparison. (Adapted from Murray, 1967.)

Motion in the frontal and transverse planes during the gait cycle was studied electrogoniometrically by Johnston and Smidt (1969) (Fig. 5–5). In the frontal plane abduction occurs during swing phase and is at a maximum just after toe off; at heel strike the hip joint reverses into adduction, which continues until late stance phase. In the transverse plane, the hip joint is externally rotated throughout the swing phase, and rotates internally just before heel strike. The joint remains internally rotated until late stance phase, when external rotation again occurs. The average ranges of motion recorded for 33 normal men in this study were 12 degrees for the frontal plane and 13 degrees for the transverse plane.

With aging the gait pattern changes, and less of the range of motion is used in the joints of the lower limb. Murray et al. (1969) studied walking patterns of 67 normal men of similar weight and height ranging in age from 20 to 87 years and compared the gait patterns of older and younger men. The differences in the sagittal body positions of the older and younger men at the instant of heel strike are illustrated in Figure 5–6. The older men

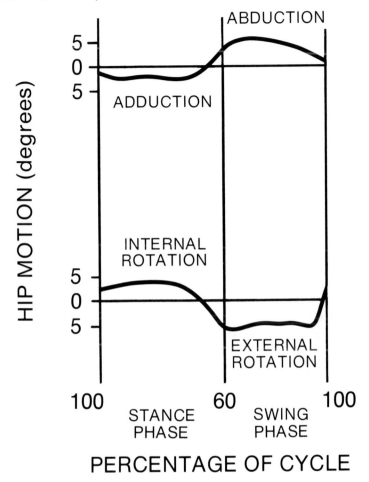

Fig. 5–5. A typical pattern for range of motion in the frontal plane (top) and transverse plane (bottom) during level walking, one gait cycle. (Adapted from Johnston and Smidt, 1969.)

showed shorter leg lengths, a decreased range of hip flexion-extension, decreased plantar flexion of the ankle, and a decreased heel-floor angle of the tracking limb; they also showed decreased dorsiflexion of the ankle and elevation of the toe of the forward limb.

The range of motion on three planes during common daily activities such as tying a shoe, sitting and rising from a chair, picking up an object from the floor, and climbing stairs was measured in 33 normal men with an electrogoniometer by Johnston and Smidt (1970). The mean motion during these activities is shown in Table 5–1. Maximal motion in the sagittal plane (hip flexion) was needed for tying the shoe and for bending down to a squatting position to pick up an object from the floor. The greatest motion in

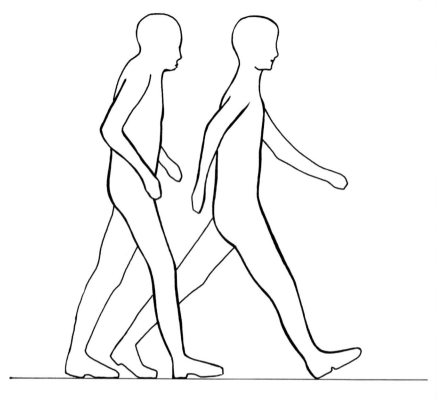

Fig. 5-6. Differences in the sagittal body positions of older men (left) and younger men (right) at the instant of heel strike. The older men show shorter leg lengths, a decreased range of hip flexion-extension, decreased plantar flexion of the ankle, and a decreased heel-floor angle of the tracking limb; they also show decreased dorsiflexion of the ankle and elevation of the toe of the forward limb. (Reprinted with permission from Murray, M. P., et al.: Walking patterns in healthy old men. J. Gerontol., 24:169–178, 1969.)

the transverse and frontal planes was recorded during tying the shoe with the foot across the opposite thigh.

The values obtained for these activities indicate that hip flexion of at least 120 degrees, abduction of at least 20 degrees, and external rotation of at least 20 degrees are necessary for carrying out daily activities in a normal manner.

Surface Joint Motion

Surface motion in the hip joint can be considered as sliding of the femoral head on the acetabulum. The pivoting of the ball and socket in three planes around the center of rotation of the femoral head produces this sliding of the joint surfaces. If incongruity is present in the femoral head, sliding may not occur parallel or tangential to the surface, and the joint cartilage may be

TABLE 5–1

MEAN MEASURES OF MAXIMUM HIP MOTION IN THREE PLANES DURING COMMON ACTIVITIES

Activity	Plane of Motion	Recorded Value (degrees)
Tying shoe with foot on floor	Sagittal	124
	Frontal	19
	Transverse	15
Tying shoe with foot across opposite thigh	Sagittal	110
	Frontal	23
	Transverse	33
Sitting down on chair and rising from sitting	Sagittal	104
	Frontal	20
	Transverse	17
Stooping to obtain object from floor	Sagittal	117
	Frontal	21
	Transverse	18
Squatting	Sagittal	122
	Frontal	28
	Transverse	26
Ascending stairs	Sagittal	67
	Frontal	16
	Transverse	18
Descending stairs	Sagittal	36

Source: Data from Johnston and Smidt, 1970. Mean for 33 normal men.

abnormally compressed or distracted. Instant center analysis cannot accurately be performed in the hip joint because motion takes place in three planes simultaneously.

KINETICS

Knowledge of the loads acting on the hip joint is important in the management of patients with hip disorders. Large forces act on the joint during simple activities. The factors involved in producing these forces must be understood if a rational rehabilitation program is to be developed for a patient with a pathologic condition of the hip.

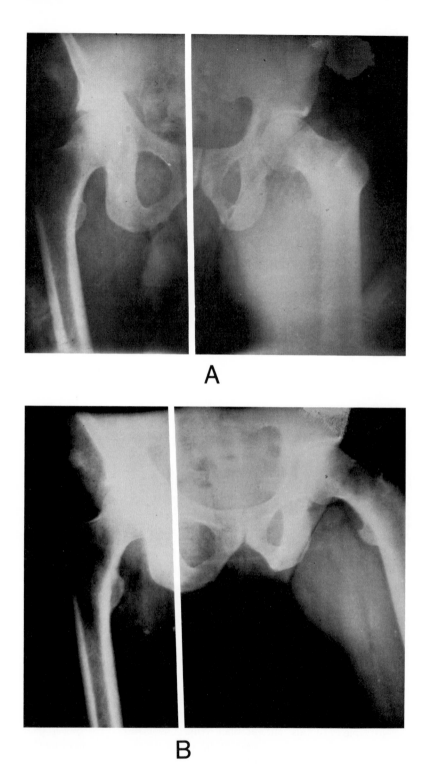

Fig. 5–7. Roentgenograms utilizing a plumb line show that the line of gravity shifts in the frontal plane with different positions of the upper body and inclinations of the pelvis. **A.** Pelvis in a neutral position. **B.** Maximum tilt of the shoulders over supporting hip joint.

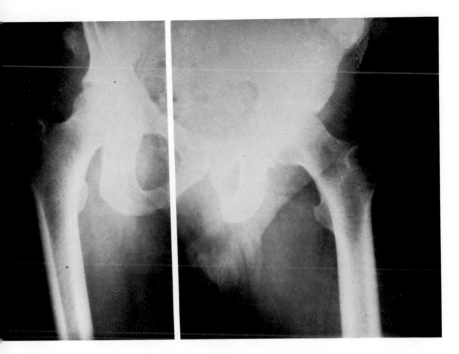

C

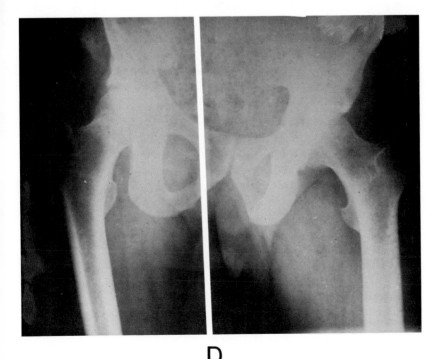

D

C. Maximum tilt of the shoulders away from the supporting hip joint.
D. Pelvis sagging away from the supporting hip joint (Trendelenburg's test). (Courtesy of John C. Baker, M.D.)

Statics

In a two-leg stance position the line of gravity passes posterior to the pubic symphysis, and since the hip joint is stable, an erect stance can be achieved without muscle contraction through the stabilizing effect of the joint capsule and capsular ligaments. Thus, the joint reaction force on the head of the femur during upright standing is one-half the superincumbent body weight. Since each lower extremity is one-sixth of body weight, the load on each hip joint will be one half of the remaining two-thirds of body weight, or one-third of body weight. If the muscles surrounding the hip joint contract to prevent swaying and to maintain an upright position of the body, this force of one-third body weight will increase in proportion to the amount of muscle activity.

In a single-leg stance the line of gravity of the superincumbent body shifts in all three planes, producing moments around the hip joint which increase the joint reaction force. The magnitude of the moments depends on the posture of the spine, the position of the non-weight-bearing leg and upper extremities, and, in particular, the inclination of the pelvis (McLeish and Charnley, 1970). Figure 5–7 demonstrates how the line of gravity in the frontal plane shifts with four different positions of the upper body and inclinations of the pelvis: standing with the pelvis in a neutral position (Fig. 5–7A), standing with a maximum tilt of the upper body over the supporting hip joint (Fig. 5–7B), standing with the upper body tilting away from the supporting hip joint (Fig. 5–7C), and standing with the pelvis sagging away from the supporting hip joint (Trendelenburg's test) (Fig. 5–7D). The joint reaction force is minimized by tilting the trunk over the hip joint.

Two methods will be illustrated for deriving the joint reaction force acting on the head of the femur. These methods are designated as the simplified free body technique for coplanar forces and the moment method utilizing equilibrium equations.

Simplified Free Body Technique for Coplanar Forces

The simplified free body technique for coplanar forces has been described in detail in Chapter 4. In the following example this technique is used to estimate the joint reaction force in the frontal plane on the femoral head during a single-leg stance with the pelvis in a neutral position. The stance limb is considered as a free body, and a free body diagram is drawn. From all the forces acting on the free body, the three main coplanar forces are identified as the force of gravity (ground reaction force) against the foot, which is transmitted through the tibia to the femoral condyles; the force produced by contraction of the abductor muscles; and the joint reaction force on the head of the femur.

—The ground reaction force (force W) has a known magnitude equal to five-sixths of body weight, and a known sense, line of application, and point of application.

—The abductor muscle force (force M) has a known sense, a known line of application and point of application estimated from the muscle origin and insertion on roentgenogram, but an unknown magnitude. Because several muscles are involved in the action of hip abduction, simplifying assumptions are made in determining the direction of this force (McLeish and Charnley, 1970). Furthermore, forces produced by other muscles active in stabilizing the hip joint are not taken into account.

—The joint reaction force (force J) has a known point of application on the surface of the femoral head, but an unknown magnitude, sense, and line of application.

The magnitudes of the abductor muscle force and the joint reaction force can be derived by designating all three forces on the free body diagram (Fig. 5–8A) and constructing a triangle of forces (Fig. 5–8B). The muscle force is

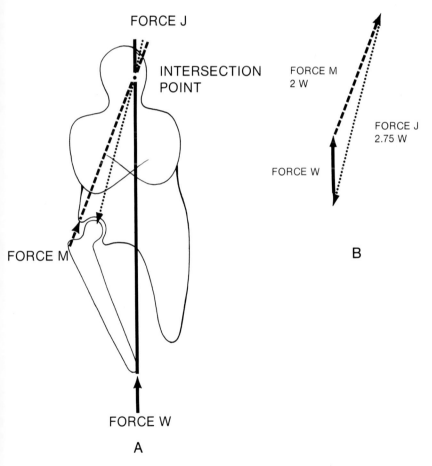

Fig. 5–8. A. On a free body diagram of the upper body and supporting lower extremity, the lines of application for forces W and M are extended until they meet (intersection point). The line of application for force J is then determined by connecting its point of application (contact point of the acetabulum and femoral head) with the intersection point for forces W and M.

    B. A triangle of forces is constructed. The magnitudes of forces M and J can be scaled from force W, which is equal to body weight. Force M is approximately two times body weight, and force J is about 2.75 times body weight.

found to be approximately two times body weight, while the joint reaction force is somewhat higher.

Moment Method Utilizing Equilibrium Equations

In the moment method, the moments acting on the joint are calculated using an equilibrium equation, and this knowledge is then utilized to calculate the joint reaction force on the head of the femur. Since in a static situation both the sum of the moments and the sum of the forces must equal zero, it is possible to derive the joint reaction force on the femoral head by finding the vertical and horizontal components of all the forces acting on the lower limb and adding these vector components.

The moment method of calculating the joint reaction force on the femoral head will be demonstrated for a single-leg stance with the pelvis level. First, the external forces acting on the body during the single-leg stance are indicated on a free body diagram (Fig. 5–9A). Since the body is in equilibrium, the ground reaction force is equal to the gravitational force of the body, which can be divided into two components: (1) the gravitational force of the stance leg, equal to one-sixth body weight; and (2) the remaining force, equal to five-sixths body weight.

Next, the body is divided at the hip joint into two free bodies. All forces and moments acting on these free bodies must be determined. The upper free body is considered first (Fig. 5–9B). In this free body two moments are required for stability. The moment arising from the superincumbent body weight (equal to five-sixths body weight) must be balanced by a moment arising from the force of the abductor muscles. The force produced by the superincumbent body weight (5/6 W) acts at a distance of b from the center of rotation of the hip (Q), thus producing a moment of 5/6 W times b. The

Fig. 5–9. **A.** External forces acting on the body during a single-leg stance with the body in equilibrium. The ground reaction force is equal to body weight (W). The gravitational force of the stance leg is equal to one-sixth of body weight; the remaining force is equal to five-sixths body weight.

    B. To find the internal forces acting on the hip joint, the joint is separated into an upper and lower free body, and the upper free body is considered first. Equilibrium is attained by the production of two equal moments. A moment arising from the force of the abductor muscle (M times c) counterbalances the moment arising from the gravitational force of the superincumbent body (5/6 W times b), which tends to tilt the pelvis away from the supporting lower extremity. Q = center of rotation of femoral head; M = abductor muscle force; 5/6 W = gravitational force of body above hip joint; c = abductor force lever arm; b = gravitational force lever arm.

    C. The muscle force (M) is equal to two times body weight and has a direction of 30 degrees from the vertical. The magnitudes of its horizontal (M_X) and vertical (M_Y) components are found by vector analysis. Perpendiculars are drawn from the tip of force M to a horizontal and a vertical line extended from the base of the force, representing M_X and M_Y, respectively. M_X and M_Y can then be scaled off. Alternatively, trigonometry can be used to find the magnitudes of the components.

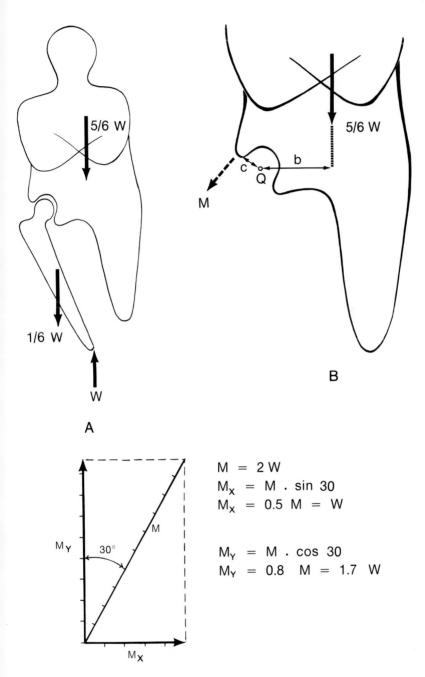

5/6 W

1/6 W

W

A

5/6 W

b

c Q

M

B

M = 2 W

$M_X = M \cdot \sin 30$

$M_X = 0.5 \; M = W$

$M_Y = M \cdot \cos 30$

$M_Y = 0.8 \quad M = 1.7 \; W$

M_Y

30°

M

M_X

C

gluteus medius muscle force (M), which acts at a distance of c from the center of rotation, produces a counterbalancing moment of M times c. For the body to remain in equilibrium, the sum of the moments must equal zero. In this example, the moments acting clockwise are considered to be positive, while the counterclockwise moments are considered to be negative. Thus,

$$(5/6 \text{ W} \times \text{b}) - (\text{M} \times \text{c}) = 0$$

$$\text{M} = 5/6 \text{ W} \times \frac{\text{b}}{\text{c}} .$$

To solve for M it is necessary to find the values of b and c. The gravitational force lever arm (b) is found roentgenographically. Since the center of gravity must lie over the base of support, a plumb line intersecting the heel can be extended upwards; a perpendicular drawn from the center of rotation of the femoral head (Q) to the line will represent distance b. The muscle force lever arm (c) is similarly found by identifying the gluteus medius muscle on roentgenogram and drawing a perpendicular from the center of rotation of the femoral head to a line approximating the muscle belly (see Fig. 5–7).

In this example a value for force M of two times body weight has been chosen. The direction of force M is found from a roentgenogram to be 30 degrees to the vertical. The horizontal and vertical components of this force are found by vector analysis (Fig. 5–9C). The horizontal component (M_x) is found to equal approximately 1.7 times body weight; the vertical component (M_y) is equal to body weight.

Attention is then directed to the lower free body (Fig. 5–10A). The gravitational forces (W and 1/6 W) are known. The joint reaction force (force J) has an unknown magnitude and direction, but must pass through the center of rotation of the femoral head. The magnitude of force J is determined by finding all the horizontal and vertical force components and summing these up (Fig. 5–10B). The vertical and horizontal components of all the forces acting on the lower free body are identified. Since the body is in equilibrium, the sum of the forces in the horizontal direction must equal

Fig. 5–10. **A.** The supporting lower extremity is considered as a free body, and the forces acting on the free body are identified. M = abductor muscle force; J = joint reaction force; 1/6 W = gravitational force of the limb; W = ground reaction force; Q = center of rotation of the femoral head.

B. The forces acting on the lower free body are divided into horizontal and vertical components. The magnitudes of J_Y and J_X are found from equilibrium equations.

C. Addition of the horizontal and vertical components J_X and J_Y is performed graphically and the joint reaction force (J) is scaled off. Its direction is measured on the parallelogram of forces.

D. The joint reaction force has a magnitude of approximately 2.7 times body weight and acts at an angle of 69 degrees from the horizontal.

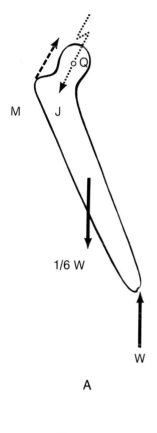

A

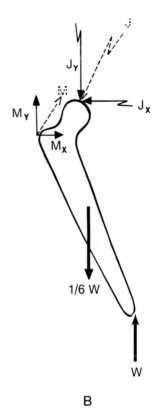

B

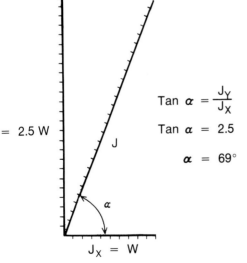

$$\text{Tan } \alpha = \frac{J_Y}{J_X}$$

$$\text{Tan } \alpha = 2.5$$

$$\alpha = 69°$$

C

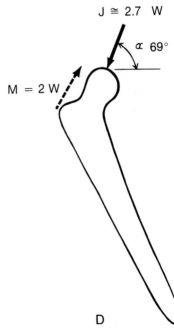

D

zero, and the sum of the forces in the vertical direction must equal zero. The horizontal and vertical forces are added:

$$M_X - J_X = 0 \qquad\qquad M_Y - J_Y - 1/6\ W + W = 0$$
$$M_X = J_X \qquad\qquad M_Y \approx 1.7\ W$$
$$M_X = W \qquad\qquad J_Y \approx 1.7\ W + 5/6\ W$$
$$J_X = W \qquad\qquad J_Y \approx 2.5\ W$$

The value of J is found by vector addition (Fig. 5–10C), and its direction is measured on the parallelogram of forces. The joint reaction force on the

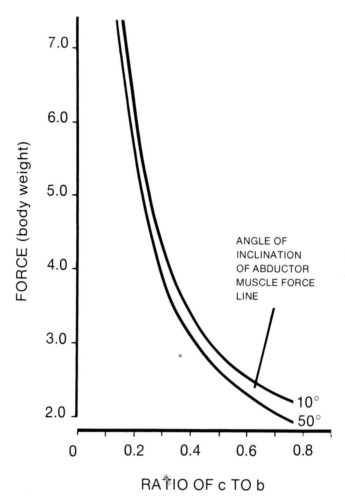

Fig. 5–11. The value of the ratio of the abductor muscle force lever arm (c) to the gravitational force lever arm (b) plotted against the joint reaction force on the femoral head in terms of body weight. Since the direction of the abductor muscle force has finite upper and lower limits (ten to 50 degrees), the force envelope is plotted. This curve can be utilized to determine the minimal force acting on the femoral head during a one-leg stance if the ratio of c to b is known. (Adapted from Frankel, 1973.)

femoral head in a single-leg stance with the pelvis level is found to be approximately 2.75 times body weight, and its direction is 69 degrees from the horizontal (Fig. 5–10D).

A key factor influencing the magnitude of the joint reaction force on the femoral head is the ratio of the abductor muscle force lever arm to the gravitational force lever arm. Figure 5–11 illustrates the relationship of this ratio to the joint reaction force. A small ratio yields a higher joint reaction force than a large ratio. A short lever arm of the abductor muscle, as in coxa valga, will result in a small ratio and thus a somewhat elevated joint reaction force. Moving the greater trochanter laterally during total hip replacement will lower the joint reaction force, as it increases the lever arm ratio. Inserting a prosthetic cup deeper in the acetabulum will reduce the gravitational lever arm, thereby increasing the ratio and decreasing the joint reaction force. It is difficult, however, to change the lever arm ratio in such a way as to greatly reduce the joint reaction force on the femoral head, as the curve becomes asymptotic when the ratio of c to b approaches 0.8.

The ratio of the abductor muscle force lever arm (c) to the gravitational force lever arm (b) was determined during a one-leg stance for a patient with

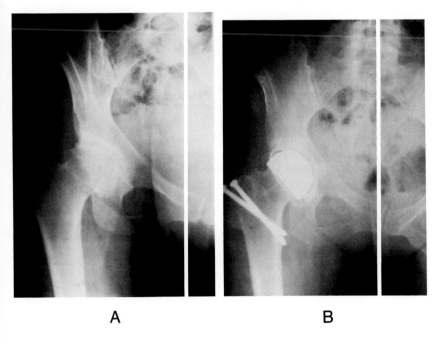

A   B

Fig. 5–12. Roentgenograms of the right hip of a patient with degenerative arthritis before and after total hip replacement. The patient stood on the right foot while lifting the left foot 3.5 cm from the floor.

 A. Before surgery. The line of gravity is represented by a plumb line.

 B. After surgery the line of gravity shifted laterally away from the affected hip joint as the patient assumed a more normal body configuration. (Courtesy of William G. Boettcher, M.D.)

degenerative arthritis of the right hip before and after total hip replacement (Fig. 5–12). After total hip replacement the line of gravity for the superincumbent body shifted laterally away from the affected hip joint as the patient assumed a more normal body configuration, having been relieved of hip joint pain by the prosthetic replacement. This shift in the line of gravity increased the gravitational force lever arm, thereby reducing the ratio of c to b. Thus, the patient was able to tolerate a higher joint reaction force on the femoral head. With the direction of the muscle force taken to be 15 degrees to the vertical, the ratio of c to b before surgery was found to be 7.0 and the joint reaction force about 2.2 times body weight. After surgery the ratio was 6.0 and the joint reaction force about 2.5 times body weight.

Dynamics

The loads on the hip joint during dynamic activities have been studied by several investigators (Paul, 1967; Rydell, 1965; Seireg and Arvikar, 1975). Utilizing a force plate system and kinematic data for the normal hip, Paul (1967) studied the joint reaction force on the femoral head in normal men and women during gait and correlated the peak magnitudes with specific muscle activity recorded electromyographically. In the men two peak forces were produced during the stance phase when the abductor muscles contracted to stabilize the pelvis. One peak of about four times body weight occurred just after heel strike, and a larger peak of about seven times body weight was reached just before toe off (Fig. 5–13A). During foot flat, the joint reaction force decreased to less than body weight because of the rapid lowering of the center of gravity of the body. During the swing phase the joint reaction force was produced by contraction of the extensor muscles in decelerating the thigh, and the magnitude remained relatively low, about equal to body weight.

In the women the force pattern was the same, but the magnitude was somewhat lower, reaching a maximum of only about four times body weight at late stance phase (Fig. 5–13B). The lower magnitude of the joint reaction force in the women may have been due to several factors: a wider female pelvis, a difference in the inclination of the femoral neck-shaft angle, a difference in footwear, and differences in the general pattern of gait.

Intravital measurements by Rydell (1965) utilizing an instrumented prosthesis also demonstrated that a large joint reaction force may act on the femoral head during the stance phase of gait. In addition, the study showed that at a faster cadence the forces acting on the prosthesis greatly increased because of an increase in the muscle activity (Fig. 5–14). The magnitude of the forces during swing phase was about half that of the forces during stance phase.

An instrumented nail plate was used to determine the forces acting on this implant during the activities of daily living following osteotomy or fracture of the femoral neck (Fig. 5–15) (Frankel et al., 1971; Lygre, 1970; Milde,

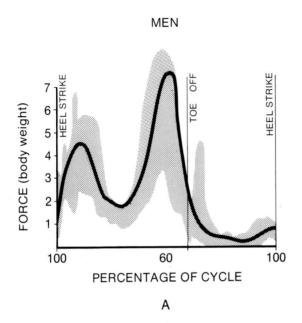

MEN

A

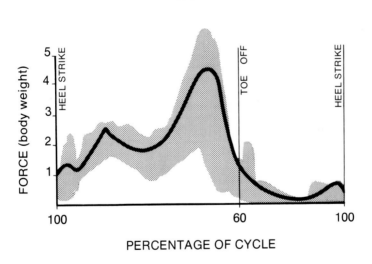

WOMEN

B

Fig. 5–13. Hip joint reaction force in terms of body weight during walking, one gait cycle. Shaded area indicates variations among subjects.
A. Force pattern for normal men.
B. Force pattern for normal women. (Adapted from Paul, 1967.)

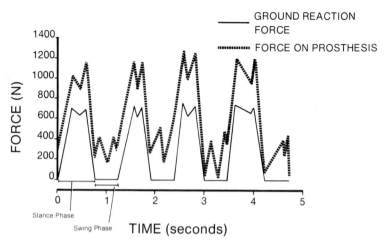

WALKING (0.9 m/s)

A

WALKING (1.3 m/s)

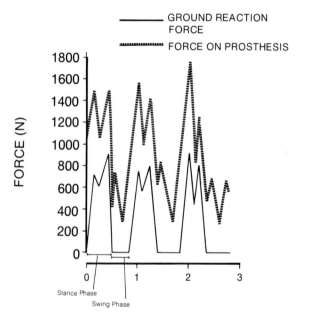

B

Fig. 5–14.   Forces on an instrumented prosthesis during walking. The broken line represents
the force on the prosthesis; the solid line represents the ground reaction force.
    A. Walking speed 0.9 meters per second.
    B. Walking speed 1.3 meters per second. (Adapted from Rydell, 1965.)

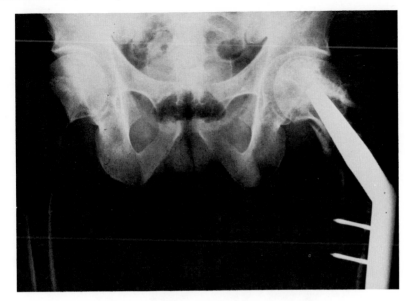

Fig. 5–15. Instrumented nail plate in the upper end of the femur used to determine the forces acting on the implant during the activities of daily living following fracture of the femoral neck. In this case the nail plate was found to transmit one-fourth of the total load on the hip joint.

1974). Although the device measured forces on the implant and not on the hip joint, it was possible to determine the proportion of the load transmitted through the device and to find the total load acting on the hip joint by static analysis. In the case shown in Figure 5–15, the nail plate transmitted one-fourth of the total load.

Large forces acting on the device were encountered during such diverse activities as raising up onto a bedpan, transferring to a wheel chair, and walking. It was found that the magnitude of the forces could be greatly modified by the skillful handling of the nurse or therapist and control by the patient. Forces of up to four times body weight were found to act on the hip joint when the patient used the elbows and heels to elevate the hips while being placed on a bedpan (Fig. 5–16A). These forces were found to be greatly reduced through the use of a trapeze and an assist from an attendant (Fig. 5–16B). The use of a hip spica cast reduced the forces acting on the hip by about two-thirds for all activities, but because the spica cast could not prevent muscle contraction during the activities, some force was still produced.

Exercises of the foot and ankle were found to produce increased forces on the head of the femur. A 5-kg extension traction on the hip had little effect in modifying the forces acting on the hip joint.

Use of the instrumented nail demonstrated that for a bedridden patient with a fractured femoral neck, the magnitude of the forces on the femoral

FORCE ON TIP OF NAIL WHEN HIPS
ARE RAISED BY PUSHING HEEL INTO BED

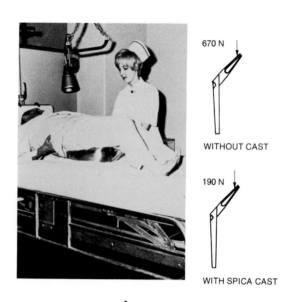

670 N

WITHOUT CAST

190 N

WITH SPICA CAST

A

FORCE ON TIP OF NAIL
DURING TRAPEZE MANEUVER TO RAISE HIPS

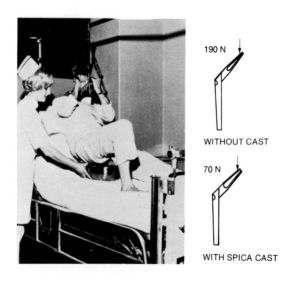

190 N

WITHOUT CAST

70 N

WITH SPICA CAST

B

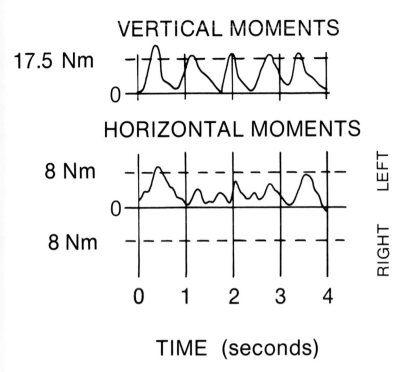

Fig. 5–17. Moments acting on the nail-plate junction in the vertical and horizontal planes while the patient walked independently.

head during the activities of daily living approach the values for the forces acting on the device during walking with external supports.

The magnitude of the moments acting on the nail-plate junction in the horizontal plane was found to be approximately one-half the magnitude of the moments acting in the vertical plane during many activities (Fig. 5–17).

Effect of External Support on the Hip Joint Reaction Force

Static analysis of the joint reaction force on the femoral head when a cane is used demonstrates that the cane should be used on the opposite side of the

Fig. 5–16. **A.** When the patient used elbows and heels to elevate the hips while being placed on a bedpan, the force on the tip on the instrumented nail was 670 newtons. With a spica cast the force on the tip of the nail was 190 newtons.

    B. The use of a trapeze and an assist from an attendant reduced the force to 190 newtons without a cast, and to 70 newtons with a spica cast. (Reprinted with permission from Frankel, V. H.: Biomechanics of the hip. In *Surgery of the Hip Joint.* Edited by R. G. Tronzo. Philadelphia, Lea & Febiger, 1973, pp. 105–125.)

painful or operated hip (Pauwels, 1936; Blount, 1956; Denham, 1959).
Using a cane on the opposite side of the painful hip reduces the force on the
femoral head of the painful joint without necessitating an antalgic body
position. The support offered by the cane greatly lessens the amount of
contraction of the abductor muscles needed to support the body weight.
Because a cane used on the opposite side of the painful hip works through a
large lever arm, a moderate force on the cane greatly decreases the abductor
muscle force, and consequently the joint reaction force, on the painful hip.
A cane used on the same side as the painful hip works through a shorter
lever arm, and thus a greater push on the cane is needed to decrease the
joint reaction force.

The use of a brace on the leg may alter the forces on the hip joint but may
not always reduce the joint reaction force on the femoral head. An ischial
long-leg brace utilized in the treatment of Perthes' disease raises the joint
reaction force during late swing phase because the large mass moment of
inertia of the brace results in a higher extensor muscle force during this part
of the gait cycle.

With the use of kinematic data from stroboscopic studies, the joint
reaction force acting on the femoral head in late swing phase of the gait
cycle was determined for an 8-year-old boy weighing 24 kg and wearing a
long-leg brace. The main muscle force (M) was produced by contraction of
the gluteus maximus muscle.

The torque about the hip joint was calculated according to the formula

$$T = I\alpha,$$

where T is the torque expressed in newton meters

    I   is the mass moment of inertia expressed in newton meters times
        seconds squared (Mm sec²)

    α is the angular acceleration in late swing phase, expressed in radians
        per second squared (r/sec¹).

In the case of the braced side,

$$I = I_L + I_B,$$

where I_L is the mass moment of inertia of the leg and I_B is the mass moment
of inertia of the brace.

On the normal side,

I = 0.45 Nm sec²

α = 24 r/sec².

Thus,

T = 0.45 Nm sec² × 24 r/sec²

T = 10.8 Nm.

On the braced side,

I = 0.45 Nm sec² + 0.35 Nm sec²

α = 24 r/sec.²

Thus,

T = (.45 Nm sec² + .35 Nm sec²)
    × 24 r/sec²

T = 19.2 Nm.

The extensor muscle force (M) was then found from the moment
relationship

$$T = Fd,$$

where F is the extensor muscle force

d is the perpendicular distance from the center of rotation of the femur to the middle of the gluteus maximus muscle.

Distance d was measured from a roentgenogram and found to be 3.2 cm. From the equation $M = \dfrac{T}{d}$, the muscle force on the normal side was calculated to be 338 newtons, and on the braced side, 600 newtons.

The joint reaction force on the femoral head (J) is equal to the muscle force (M) minus the gravitational force produced by the weight of the limb (W_L). In this example W_L was estimated to be 40 newtons.

On the normal side,

$J = M - W_L$

$J = 338\ N - 40\ N$

$J = 298\ N.$

On the braced side,

$J = M - W_L$

$J = 600\ N - 40\ N$

$J = 560\ N.$

Thus, the joint reaction force on the femoral head in the braced limb was over 50% higher than the force in the nonbraced limb, reaching more than two times body weight.

SUMMARY

1. The hip joint is a ball-and-socket joint composed of the acetabulum and femoral head.
2. The thickness and mechanical properties of the cartilage on the femoral head and acetabulum vary from point to point.
3. Hip flexion of at least 120 degrees, abduction of at least 20 degrees, and external rotation of at least 20 degrees are necessary for carrying out daily activities in a normal manner.
4. A joint reaction force of approximately three times body weight acts on the hip joint during a single-leg stance with the pelvis in a neutral position; its magnitude varies with a change in position of the upper body.
5. The magnitude of the hip joint reaction force is influenced by the ratio of the abductor muscle force and gravitational force lever arms. A small ratio yields a higher joint reaction force than a large ratio.
6. The hip joint reaction force during gait reaches levels of six times body weight or more in stance phase and is approximately equal to body weight during swing phase.
7. An increase in gait velocity increases the magnitude of the hip joint reaction force in both swing and stance phase.
8. The forces acting on an internal fixation device during the activities of daily living vary greatly depending on the nursing and therapeutic activities undergone by the patient.

9. The use of a brace on the leg can alter the magnitude of the hip joint reaction force.

REFERENCES

Blount, W. P.: Don't throw away the cane. J. Bone Joint Surg., *38A*:695, 1956.

Denham, R. A.: Hip mechanics. J. Bone Joint Surg., *41B*:550, 1959.

Frankel, V. H.: *The Femoral Neck: Function, Fracture Mechanisms, Internal Fixation.* Springfield, Charles C Thomas, 1960.

Frankel, V. H.: Biomechanics of the hip. In *Surgery of the Hip Joint.* Edited by R. G. Tronzo. Philadelphia, Lea & Febiger, 1973, pp. 105–125.

Frankel, V. H., Burstein, A. H., Lygre, L., and Brown, R. H.: The telltale nail. In Proceedings of the American Academy of Orthopaedic Surgeons, Scientific Exhibits—1971. J. Bone Joint Surg., *53A*:1232, 1971.

Inman, V. T.: Functional aspects of the abductor muscles of the hip. J. Bone Joint Surg., *29*:607, 1947.

Johnston, R. C., and Smidt, G. L.: Measurement of hip-joint motion during walking. Evaluation of an electrogoniometric method. J. Bone Joint Surg., *51A*:1083, 1969.

————: Hip motion measurements for selected activities of daily living. Clin. Orthop., *72*:205, 1970.

Kempson, G. E., Spivey, C. J., Swanson, S. A. V., and Freeman, M. A. R.: Patterns of cartilage stiffness on normal and degenerate human femoral heads. J. Biomech., *4*:597, 1971.

Lygre, L.: The loads produced on the hip joint by nursing procedures: A telemeterization study. M.S. thesis, Case Western Reserve University, 1970.

McLeish, R. D., and Charnley, J.: Abduction forces in the one-legged stance. J. Biomech., *3*:191, 1970.

Milde, F. K.: Loads on femoral head during nursing care activities as measured by a telemeterized nail-plate. M.S. thesis, Case Western Reserve University, 1974.

Murray, M. P.: Gait as a total pattern of movement. Am. J. Phys. Med., *46*:290, 1967.

Murray, M. P., Kory, R. C., and Clarkson, B. H.: Walking patterns in healthy old men. J. Gerontol., *24*:169, 1969.

Paul, J. P.: Forces at the human hip joint. Ph.D. thesis, University of Chicago, 1967.

Pauwels, F.: *Der Schenkelhalsbruch, ein mechanisches Problem: Grundlagen des heilungsvorganges Prognose und kausale Therapie.* Stuttgart, Ferdinand Enke, 1936.

Rydell, N.: Forces in the hip-joint, part (II) intravital measurements. In *Biomechanics and Related Bio-Engineering Topics.* Edited by R. M. Kenedi. Oxford and Edinburgh, Pergamon Press, 1965, pp. 351–357.

Seireg, A., and Arvikar, R. J.: The prediction of muscular load sharing and joint forces in the lower extremities during walking. J. Biomech. *8*:89, 1975.

SUGGESTED READING

Blount, W. P.: Don't throw away the cane. J. Bone Joint Surg., *38A*:695–708, 1956.

Charnley, J., and Pusso, R.: The recording and analysis of gait in relation to the surgery of the hip joint. Clin. Orthop., *58*:153–164, 1968.

Denham, R. A.: Hip mechanics. J. Bone Joint Surg., *41B*:550–557, 1959.

Frankel, V. H.: *The Femoral Neck: Function, Fracture Mechanisms, Internal Fixation.* Springfield, Charles C Thomas, 1960.

Frankel, V. H.: Biomechanics of the hip. In *Surgery of the Hip Joint.* Edited by R. G. Tronzo. Philadelphia, Lea & Febiger, 1973, pp. 105–125.

Frankel, V. H., Burstein, A. H., Lygre, L., and Brown, R. H.: The telltale nail. In Proceedings of the American Academy of Orthopaedic Surgeons. Scientific Exhibits—1971. J. Bone Joint Surg., *53A*:1232, 1971.

Gore, D. R., Murray, M. P., Sepic, S. B., and Gardner, G. M.: Walking patterns of men with unilateral surgical hip fusion. J. Bone Joint Surg., *57A*:759–765, 1975.

Inman, V. T.: Functional aspects of the abductor muscles of the hip. J. Bone Joint Surg., *29*:607–619, 1947.

Johnston, R. C.: Detailed analysis of hip joint during gait. In *The Hip. Proceedings of the Second Open Scientific Meeting of the Hip Society, 1974.* Edited by W. H. Harris. Saint Louis, C. V. Mosby Co., 1974, pp. 94–110.

Johnson, R. C., and Smidt, G. L.: Measurement of hip-joint motion during walking. Evaluation of an electrogoniometric method. J. Bone Joint Surg., *51A*:1083–1094, 1969.

———: Hip motion measurements for selected activities of daily living. Clin. Orthop., *72*:205–215, 1970.

Kempson, G. E., Spivey, C. J., Swanson, S. A. V., and Freeman, M. A. R.: Patterns of cartilage stiffness on normal and degenerate human femoral heads. J. Biomech., *4*:597–609, 1971.

Lygre, L.: The loads produced on the hip joint by nursing procedures: A telemeterization study. M.S. thesis, Case Western Reserve University, 1970.

McLeish, R. D., and Charnley, J.: Abduction forces in the one-legged stance. J. Biomech., *3*:191–209, 1970.

Merchant, A. C.: Hip abductor muscle force. An experimental study of the influence of hip position with particular reference to rotation. J. Bone Joint Surg., *47A*:462–476, 1965.

Milde, F. K.: Loads on femoral head during nursing care activities as measured by a telemeterized nail-plate. M.S. thesis, Case Western Reserve University, 1974.

Murray, M. P.: Gait as a total pattern of movement. Am. J. Phys. Med., *46*:290–333, 1967.

Murray, M. P., and Peterson, R. M.: Weight distribution and weight-shifting activity during normal standing posture. Phys. Ther., *53*:741–748, 1973.

Murray, M. P., and Sepic, S. B.: Maximum isometric torque of hip abductor and adductor muscles. J. Am. Phys. Ther. Assoc., *48*:1327–1335, 1968.

Murray, M. P., Gore, D. R., and Clarkson, B. H.: Walking patterns of patients with unilateral hip pain due to osteo-arthritis and avascular necrosis. J. Bone Joint Surg., *53A*:259–274, 1971.

Murray, M. P., Kory, R. C., and Clarkson, B. H.: Walking patterns in healthy old men. J. Gerontol., *24*:169–178, 1969.

Murray, M. P., Seireg, A. A., and Scholz, R. C.: Center of gravity, center of pressure, and supportive forces during human activities. J. Appl. Physiol., *23*:831–838, 1967.

———: A survey of the time, magnitude and orientation of forces applied to walking sticks by disabled men. Am. J. Phys. Med., *48*:1–13, 1969.

Paul, J. P.: Forces at the hip joint. Ph.D. thesis, University of Chicago, 1967.

Pauwels, F.: *Der Schenkelhalsbruch, ein mechanisches Problem: Grundlagen des heilungsvorganges Prognose und kausale Therapie.* Stuttgart, Ferdinand Enke, 1936.

Rydell, N.: Forces in the hip-joint, part (II) intravital measurements. In *Biomechanics and Related Bio-Engineering Topics.* Edited by R. M. Kenedi. Oxford and Edinburgh, Pergamon Press, 1965, pp. 351–357.

Seireg, A., and Arvikar, R. J.: The prediction of muscular load sharing and joint forces in the lower extremities during walking. J. Biomech. *8*:89–102, 1975.

CHAPTER 6

Biomechanics of the Ankle

Victor H. Frankel
and
Margareta Nordin

Like the other major joints in the lower extremity, the ankle joint participates in kinematic functions and load bearing. This joint consists of the tibiotalar, fibulotalar, and distal tibiofibular joints (Fig. 6–1). Because of its anatomical configuration, the ankle joint is more like the hip joint, which is inherently stable, than the knee joint, which requires ligamentous and muscular restraints. The ankle mortise is maintained by the shape of the articulation, the lateral and medial collateral ligament system, the joint capsule, and the interosseous ligaments.

Again like the hip joint, the ankle joint responds poorly to small changes in its anatomical configuration. Loss of kinematic and structural restraints due to severe sprains can seriously affect ankle stability and can produce malalignment of the ankle joint surfaces. This malalignment may result in profound pathological changes.

KINEMATICS

The ankle joint is basically a uniplanar joint; motion of talus takes place primarily in the sagittal plane about a transverse axis that deviates posteriorly from the frontal plane (Barnett and Napier, 1952; Inman, 1976). This motion allows dorsiflexion and plantar flexion of the foot. The talus in the mortise may also rotate a few degrees around a longitudinal axis and tilt a few degrees around a sagittal axis. Deviations in the inclination of any of the axes of the ankle joint due to epiphyseal injury, ligamentous injury, or malunion of a fractured tibia can result in severe pathological changes in the joint.

Range of Motion

The total range of motion of the ankle joint in the sagittal plane is approximately 45 degrees but can vary widely among individuals and with age. Ten to 20 degrees of this motion is defined as dorsiflexion and the remaining 25 to 35 degrees as plantar flexion.

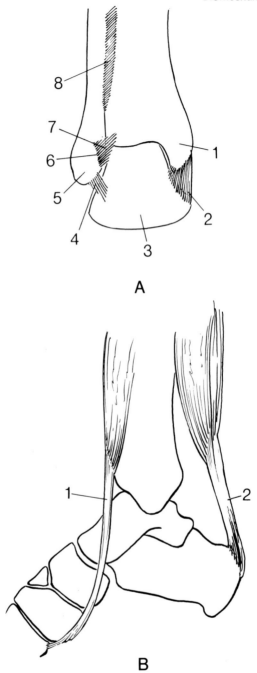

Fig. 6–1.   Ankle joint composed of the tibiotalar, fibulotalar, and distal tibiofibular joints.
  A. Front view: 1. tibia; 2. deltoid ligament; 3. talus; 4. anterior talofibular ligament; 5. fibula; 6. anterior-inferior tibiotalar ligament; 7. distal tibiofibular joint; 8. interosseous membrane.
  B. Medial view: 1. tibialis anterior tendon; 2. Achilles tendon.

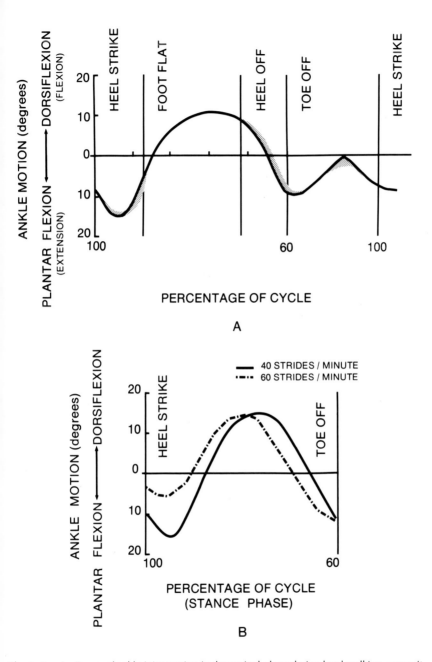

Fig. 6–2. **A.** Range of ankle joint motion in the sagittal plane during level walking, one gait cycle. Shaded area indicates variation among 60 subjects (age range—20 years to 65 years). (Adapted from Murray et al., 1964.)

   B. Range of ankle joint motion for 5 men during the stance phase of gait at two velocities. The amount of plantar flexion at heel strike decreased in the faster cadence, and the peak plantar flexion occurred earlier in the midstance phase. (Adapted from Stauffer et al., 1977.)

The normal pattern of ankle joint motion during gait has been studied extensively (Fig. 6–2A) (Murray et al., 1964; Wright et al., 1964; Lamoreaux, 1971; Stauffer et al., 1977). At heel strike the ankle is in a slightly plantar-flexed position. Further plantar flexion occurs until foot flat. The motion rapidly reverses to dorsiflexion during midstance as the body passes over the supporting foot; plantar flexion then occurs again after heel off at the end of the stance phase. At toe off at the beginning of the swing phase the ankle is plantar flexed; the motion reverses to dorsiflexion in the middle of the swing phase and again changes to slight plantar flexion at heel strike. The amount of plantar flexion at heel strike depends upon the heel height of the shoe worn. The higher the heel is, the greater the amount of plantar flexion (Stauffer et al., 1977). The total amount of motion during the gait cycle, however, decreases with an increased heel height (Murray et al., 1970).

Sammarco et al. (1973) studied total ankle joint motion roentgeno-graphically and recorded the average range of motion in the sagittal plane during gait for 24 normal subjects ranging in age from 20 to 60 years. The total range varied from 24 to 75 degrees, with an average of 43±12.7 degrees, and tended to decrease with age. The amount of dorsiflexion and plantar flexion was almost equal (21 and 23 degrees, respectively).

Stauffer et al. (1977) studied normal gait in 5 men at two velocities and found that the amount of plantar flexion at heel strike decreased with an increased velocity (Fig. 6–2B). Dorsiflexion remained essentially unchanged in the two cadences, but the peak plantar flexion occurred earlier in the midstance phase at the faster velocity. This situation contrasts with that in both the hip and knee joint, where joint motion has been found to increase directly with increased cadence (Pauwels, 1936; Rydell, 1966).

Patients with ankle joint disease have been found to have a wide variance in total ankle joint motion (Sammarco et al., 1973; Stauffer et al., 1977). Sammarco et al. (1973) measured joint motion in the sagittal plane during gait in 10 patients with abnormal ankle joints. The range of motion showed an overall decrease compared with the range in a control group. The greatest decrease took place in dorsiflexion. Stauffer et al. (1977) found the same pattern in nine patients with abnormal ankles.

Surface Joint Motion

In the ankle joint, surface motion occurs in both the tibiotalar and fibulotalar articulations. During plantar flexion a few degrees of motion in the distal tibiofibular joint takes place to accommodate the decreased width of the posterior talus. Sammarco et al. (1973) performed instant center analyses on 24 normal weight-bearing ankles and on 11 ankles undergoing disease processes. Utilizing multiple roentgenographic exposures and the instant center technique as described for the knee in Chapter 4, they determined the instant center pathway in the normal and abnormal ankles for a range of motion from full dorsiflexion to full plantar flexion. The direction of displacement of the contact points was then determined.

In the normal ankles the joint surfaces distracted at the beginning of the motion, and sliding then took place (Fig. 6–3); the motion ended with jamming of the surfaces. During the reverse motion the jammed surfaces distracted and then jammed again at the completion of motion. Sliding occurred throughout the range of normal motion. It is possible that this distraction and jamming of the joint surfaces plays an important role in lubrication of the joint.

In the abnormal ankles the direction of displacement showed no consistent pattern. Joint distraction took place in an unpredictable manner, and jamming occurred when the joint was in a neutral position rather than at the end of dorsiflexion.

Stauffer et al. (1977) also studied the instant center pattern of ankle joint motion in the sagittal plane. They found that for the range of motion used in gait the instant centers all fell well within the body of the talus, and so close

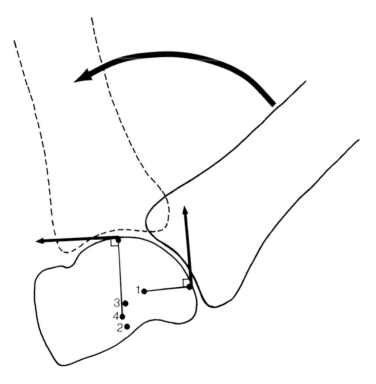

Fig. 6–3. Instant center pathway for surface joint motion in a normal ankle from full dorsiflexion to full plantar flexion. All instant centers fall within the talus. The direction of displacement of the contact points shows distraction of the joint surfaces at the beginning of the motion (point 1) and sliding thereafter (point 4). (Adapted from Sammarco et al., 1973.)

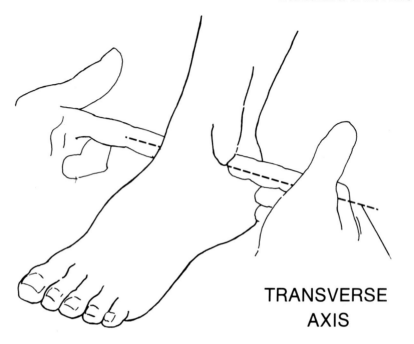

**TRANSVERSE
AXIS**

Fig. 6–4. Clinical estimation of the position of the axis of motion of the ankle joint by palpation of the malleoli. The axis deviates posteriorly from the frontal plane and is not rigid but changes slightly with dorsiflexion and plantar flexion. (Adapted from Inman, 1973.)

together that the instant center of the ankle joint during gait could be considered as a single point.

Barnett and Napier (1952) showed in a study of 152 tali that the medial and lateral profiles of the talus have different curvatures. This finding indicates that the inclination of the axis of motion of the ankle joint is not rigid but changes slightly with motion. This change in the axis was confirmed by Inman (1976) and by the instant center analyses of Sammarco et al. (1973). Clinically, the position of the axis can be estimated by palpating the malleoli (Fig. 6–4) (Inman, 1973).

Ankle Joint Stability

The mortise of the ankle joint is maintained by the shape of the talus, its tight fit between the fibula and tibia, and the interosseous membrane between these two bones. This stabilizing effect is greatest in dorsiflexion. The mortise is further maintained by the anterior and posterior talofibular ligaments and by the ankle joint capsule. The anterior talofibular ligament and the calcaneofibular ligament are important for stabilizing the lateral side of the ankle joint, while the deltoid ligament secures the medial side of the joint. Because the talus narrows posteriorly, mortise stability in plantar flexion is provided mainly by the elasticity of the ligaments. The musculotendinous apparatus surrounding the ankle joint on the medial and lateral sides plays a small role in stabilizing the joint.

KINETICS

The joint reaction force on the ankle joint during gait is equivalent to or somewhat greater than the forces previously determined for the hip and knee joints. The ankle joint, however, has a large weight-bearing surface area; thus, lower loads per unit area (stresses) are transmitted across this joint than in the hip or knee joint. The following static and dynamic analyses give an estimate of the magnitude of the joint reaction forces acting on the ankle joint during standing on tiptoe on one leg and during level walking.

Statics

When an individual stands on both feet, each ankle joint supports approximately one-half the body weight. If muscle activity is involved in balancing the body, the joint reaction force on the ankle will increase in proportion to the amount of muscle force used for these balancing activities.

In a static analysis of the forces acting on the ankle joint, the magnitude of the force produced by contraction of the gastrocnemius and the soleus muscles through the Achilles tendon, and therefore the magnitude of the joint reaction force, can be calculated by using the simplified free body technique described for the knee in Chapter 4.

In the following example, the Achilles tendon force and joint reaction force of the ankle joint are calculated for an individual standing on tiptoe on one leg. For the body to maintain equilibrium, the line of gravity for the body must pass through the ball of the foot to the sole of the shoe, and thence to the ground.

In this example the foot, including the talus, is considered as a free body. From all the forces acting on this free body the three main coplanar forces are identified as the gravitational force, the tensile force through the Achilles tendon, and the joint reaction force on the dome of the talus.

—The gravitational force (W) has a known magnitude (equal to body weight), sense, line of application, and point of application.

—The Achilles tendon force (A), which maintains the plantar-flexed position of the foot, has a known sense, line of application (along the Achilles tendon), and point of application (point of insertion of the Achilles tendon into the calcaneus), but an unknown magnitude.

—The joint reaction force (J) has a known point of application at the dome of the talus (estimated from a roentgenogram), but an unknown magnitude, sense, and line of application.

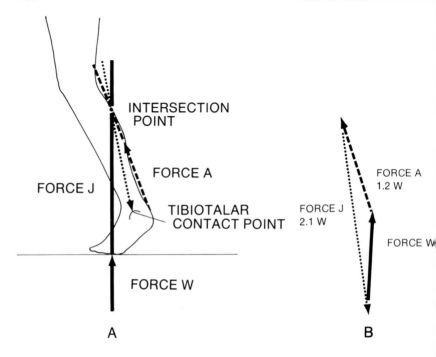

Fig. 6–5. A. On a free body diagram of the foot, including the talus, the lines of application for forces W and A are extended until they intersect (intersection point). The line of application for force J is then determined by connecting its point of application (tibiotalar contact point) with the intersection point for forces W and A.
B. A triangle of forces is constructed. Force A is 1.2 times body weight, and force J is 2.1 times body weight.

The magnitude of the Achilles tendon force (A) and the joint reaction force (J) can be derived by designating the forces on a free body diagram and constructing a triangle of forces (Fig. 6–5). Not surprisingly, these forces are found to be quite high. The joint reaction force is about 2.1 times body weight, and the Achilles tendon force reaches about 1.2 times body weight. The large amount of force required for rising up on tiptoes explains why the patient with weak gastrocnemius and soleus muscles has difficulty in rising to tiptoes ten times rapidly. It also explains why a patient with degenerative arthritis of the ankle joint will have pain on rising up on tiptoes.

Lambert (1971) studied the weight-bearing function of the fibula and demonstrated that a portion of the compressive force transmitted from the knee to the ankle is transmitted through the fibula. In a static model of the ankle joint, including the fibula, he determined that approximately one-sixth of the load of the leg is borne by the fibula and is transmitted to the fibular facet on the talus. He found that the load on the fibula originated at the proximal tibiofibular joint and that little of the load was transmitted through the interosseous membrane.

Dynamics

Dynamic studies of the ankle joint are important for understanding the magnitude of the loads on the normal ankle joint during exercise, the loads on the damaged ankle joint during normal activities, and the loads expected to act on an ankle joint prosthesis.

The loads on the ankle joint during level walking have been studied by Stauffer et al. (1977). Utilizing a force plate, high-speed photography, roentgenograms, and free body calculations, they determined both compressive and shear forces acting on the ankle joint during the stance phase of gait. These forces were calculated for normal subjects and for patients with ankle joint disease before and after prosthetic ankle replacement.

In the normal subjects the main compressive force across the ankle joint during gait was produced by contraction of the gastrocnemius and soleus muscles and was transmitted through the Achilles tendon. The force produced by contraction of the pretibial muscle group acted only during early stance phase with a low magnitude, less than 20% of body weight. The Achilles tendon force attained high levels during late stance phase when the Achilles tendon began to produce a torque for plantar flexion at push off. The highest joint compression force occurred at this point in the gait cycle: about five times body weight (Fig. 6–6A). The shear force for these subjects reached its maximum value, about 0.8 times body weight, just after the middle of the stance phase during heel off (Fig. 6–6B).

The patients with ankle joint disease showed a decreased joint compression force: about three times body weight. The peak compressive force occurred slightly earlier in these patients than in the normal subjects. The shear forces also showed a decrease. A follow-up 1 year after ankle joint replacement in these patients showed no change in the joint compression force patterns. The shear forces, however, showed a pattern and magnitude nearly equivalent to those in the normal subjects.

Stauffer et al. (1977) showed the effect of two walking cadences on the ankle joint reaction force in normal subjects. The patterns were somewhat different for the two cadences, but the magnitudes of the peak forces were the same (Fig. 6–7). In the faster cadence the pattern showed two peak forces of three to five times body weight, one in early stance phase and one in late stance phase. In the slower cadence only one peak force of approximately five times body weight was reached during late stance phase. It should be noted that these dynamic studies assume complete transmission of the force through the tibiotalar joint and do not take into account any load-bearing in this joint.

The ankle joint has a load-bearing surface of 11 to 13 square cm (Greenwald, 1977). This large weight-bearing surface can result in lower stresses across this joint than in the knee or hip joint. A small deviation in the anatomical configuration of the ankle joint can result in gross changes in the weight-bearing pattern and consequent peak loads. Ramsey and Hamilton (1976) noted changes in the tibiotalar contact area produced by lateral talar

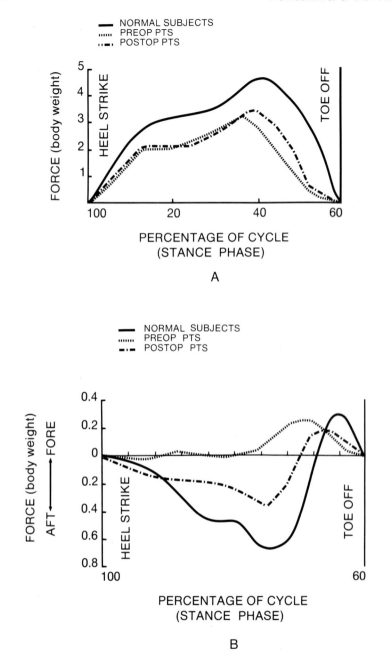

Fig. 6–6. **A.** Ankle joint compression force during the stance phase of level walking for 5 normal subjects and for 9 patients with ankle joint disease before and after prosthetic ankle replacement.

   B. Shear force (in the "fore" and "aft" directions) produced in the ankle joint during the stance phase of level walking for the same subjects. (Adapted from Stauffer et al., 1977.)

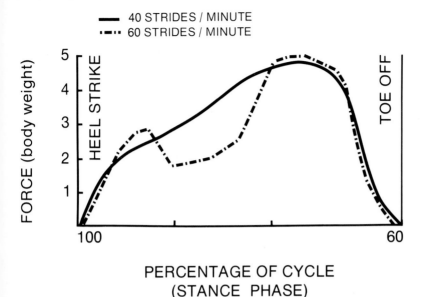

Fig. 6–7. Ankle joint reaction force in 5 normal subjects during the stance phase of gait at two velocities. Although the patterns varied somewhat for the two cadences, the magnitudes of the peak forces were the same. In the faster cadence two peak forces of three to five times body weight occurred, one in early stance phase and one in late stance phase. In the slower cadence only one peak force of approximately five times body weight was reached during late stance phase. (Adapted from Stauffer et al., 1977.)

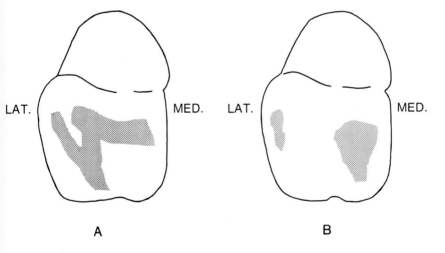

Fig. 6–8. Top view of the left talus showing the tibiotalar contact area (shaded).
   A. No talar shift (displacement). The major contact area is on the lateral side.
   B. Two millimeters of lateral talar shift. The total contact area has markedly decreased, and the major contact area is now on the medial side. (Adapted from Ramsey and Hamilton, 1976.)

shift (Fig. 6–8). Lateral talar shift occurs quite frequently following major sprains and fractures of the ankle joint, and if not corrected can lead to gross biomechanical alterations in the ankle joint. In the case of a shift of only 1 to 2 mm, the contact surface stresses would rise precipitously, and early degenerative changes in the ankle joint could result.

SUMMARY

1. The ankle joint consists of the tibiotalar, fibulotalar, and distal tibiofibular joints.
2. Motion of the talus takes place primarily in the sagittal plane about a transverse axis that deviates posteriorly from the frontal plane.
3. The instant center of the ankle joint for the range of motion used in gait can be considered as one point.
4. The forces in the ankle joint during gait can rise to levels of five times body weight or more.
5. The fibulotalar joint transmits approximately one-sixth of the force exerted through the leg.
6. Small anatomical deviations in the tibiotalar and fibulotalar articulations can result in gross changes in the magnitude and direction of the stresses on the talus.

REFERENCES

Barnett, C. H., and Napier, J. R.: The axis of rotation at the ankle joint in man. Its influence upon the form of the talus and the mobility of the fibula. J. Anat., *86*:1, 1952.
Greenwald, S.: Unpublished data. Cited in Stauffer, R. N., Chao, E. Y. S., and Brewster, R. C.: Force and motion analysis of the normal, diseased, and prosthetic ankle joint. Clin. Orthop., *127*:189, 1977.
Inman, V. T.: *DuVries' Surgery of the Foot.* St. Louis, C. V. Mosby Co., 1973.
———: *The Joints of the Ankle.* Baltimore, Williams & Wilkins, 1976.
Lambert, L. L.: The weight-bearing function of the fibula. A strain gauge study. J. Bone Joint Surg., *53A*:507, 1971.
Lamoreaux, L. W.: Kinematic measurements in the study of human walking. Bull. Prosthet. Res., *10*:3, 1971.
Murray, M. P., Drought, A. B., and Kory, R. C.: Walking patterns in normal men. J. Bone Joint Surg., *46A*:335, 1964.
Murray, M. P., Kory, R. C., and Sepic, S. B.: Walking patterns of normal women. Arch. Phys. Med. Rehabil., *51*:637, 1970.
Pauwels, F.: *Der Schenkelhalsbruch, ein mechanisches Problem: Grundlagen des heilungsvorganges Prognose und kausale Therapie.* Stuttgart, Ferdinand Enke, 1936.
Ramsey, P. L., and Hamilton, W.: Changes in tibiotalar area of contact caused by lateral talar shift. J. Bone Joint Surg., *58A*:356, 1976.
Rydell, N. W.: Forces acting on the femoral head prosthesis. A study on strain gauge supplied prostheses in living persons. Acta Orthop. Scand., Suppl. *88*, pp. 1–132, 1966.
Sammarco, G. J., Burstein, A. H., and Frankel, V. H.: Biomechanics of the ankle: A kinematic study. Orthop. Clin. North Am., *4*:75, 1973.
Stauffer, R. N., Chao, E. Y. S., and Brewster, R. C.: Force and motion analysis of the normal, diseased, and prosthetic ankle joint. Clin. Orthop., *127*:189, 1977.
Wright, D. G., Desai, S. M., and Henderson, W. H.: Action of the subtalar and ankle-joint complex during the stance phase of walking. J. Bone Joint Surg., *46A*:361, 1964.

SUGGESTED READING

Barnett, C. H., and Napier, J. R.: The axis of rotation at the ankle joint in man. Its influence upon the form of the talus and the mobility of the fibula. J. Anat., *86*:1–9, 1952.

Bresler, B., and Frankel, J. P.: The forces and moments in the leg during level walking. Trans. Am. Soc. Mech. Engng., *72*:27–36, 1950.

Close, J. R.: Some applications of the functional anatomy of the ankle joint. J. Bone Joint Surg., *38A*:761–781, 1956.

———: *Motor Function in the Lower Extremity. Analyses by Electronic Instrumentation.* American Lecture Series, No. 551. Springfield, Charles C Thomas, 1964.

Inman, V. T.: The influence of the foot-ankle complex on the proximal skeletal structures. Artif. Limbs, *13*:59–65, 1969.

———: *DuVries' Surgery of the Foot.* St. Louis, C. V. Mosby Co., 1973.

———: *The Joints of the Ankle.* Baltimore, Williams & Wilkins, 1976.

Isman, R. E., and Inman, V. T.: Anthropometric studies of the human foot and ankle. Bull. Prosthet. Res., *10–11*:97, 1969.

Kapandji, I. A.: *The Physiology of the Joints.* Vol. 2. Lower Limb. Edinburgh, London, and New York, Churchill Livingstone, 1975.

Kempson, G. E., Freeman, M. A. R., and Tuke, M. A.: Engineering considerations in the design of an ankle joint. Biomed. Engng., *10*:166–171, 180, 1975.

Lambert, L. L.: The weight-bearing function of the fibula. A strain gauge study. J. Bone Joint Surg., *53*:507–513, 1971.

Lamoreaux, L. W.: Kinematic measurements in the study of human walking. Bull. Prosthet. Res., *10*:3–84, 1971.

Miura, M., Miyashita, M., Matsui, H., and Sodeyama, H.: Photographic method of analyzing the pressure distribution of the foot against the ground. Biomech., *7*:482–487, 1974.

Murray, M. P., Drought, A. B., and Kory, R. C.: Walking patterns in normal men. J. Bone Joint Surg., *46A*:335–360, 1964.

Murray, M. P., Kory, R. C., and Sepic, S. B.: Walking patterns of normal women. Arch. Phys. Med. Rehabil., *51*:637–650, 1970.

Murray, M. P., Guten, G. N., Baldwin, J. M., and Gardner, G. M.: A comparison of plantar flexion torque with and without the triceps surae. Acta Orthop. Scand., *47*:122–124, 1976.

Murray, M. P., Guten, G. N., Sepic, S. B., Gardner, G. M., and Baldwin, J. M.: Function of the triceps surae during gait. Compensatory mechanisms for unilateral loss. J. Bone Joint Surg., *60A*:473–476, 1978.

Paul, J. P.: Bio-engineering studies of the forces transmitted by joints. II. Engineering analysis. In *Biomechanics and Related Bioengineering Topics.* Edited by R. N. Kenedi. Edinburgh, London, and New York, Pergamon Press, 1965, pp. 368–380.

Pauwels, F.: *Der Schenkelhalsbruch, ein mechanisches Problem: Grundlagen des heilungsvorganges Prognose und kausale Therapie.* Stuttgart, Ferdinand Enke, 1936.

Ramsey, P. L., and Hamilton, W.: Changes in tibiotalar area of contact caused by lateral talar shift. J. Bone Joint Surg., *58A*:356–357, 1976.

Rydell, N. W.: Forces acting on the femoral head prosthesis. A study on strain gauge supplied prostheses in living persons. Acta Orthop. Scand., Suppl. *88*:1–132, 1966.

Sammarco, G. J., Burstein, A. H., and Frankel, V. H.: Biomechanics of the ankle: A kinematic study. Orthop. Clin. North Am., *4*:75–96, 1973.

Stauffer, R. N., Chao, E. Y. S., and Brewster, R. C.: Force and motion analysis of the normal, diseased, and prosthetic ankle joint. Clin. Orthop., *127*:189–196. 1977.

Wright, D. G., Desai, S. M., and Henderson, W. H.: Action of the subtalar and ankle-joint complex during the stance phase of walking. J. Bone Joint Surg., *46A*:361–464, 1964.

CHAPTER 7

Biomechanics of the Foot

G. James Sammarco

Biomechanics of the foot is an extremely complex subject. It may be considered distinctly different from that of the ankle. However, several authors in the past have considered portions of the foot and ankle as a single unit, most notably Isman and Inman (1968). Biomechanically the foot must also be considered an integral part of the entire lower extremity.

The unique qualities that the foot possesses allow it to be rigid when necessary, converting the 26 bones into a single solid unit, or quite flexible when necessary, as during climbing barefooted. Between these two extremes of the rigid and flexible foot lies the motion of the foot during gait. The necessity for variety of activity within the structure of the foot is due to the fact that the surfaces on which we stand and move vary considerably from soft, smooth, and slippery to firm, rough, and sticky. In addition, foot coverings vary from no covering to the sophisticated modern ski boot, which is thoroughly rigid and snug-fitting below the ankle.

This chapter will discuss the motion which occurs within the foot during various phases of gait as well as at the extremes of motion. Location of forces as they pass from the tibiofibular complex into the dome of the talus, and then into the foot, will also be covered.

A discussion of sophisticated electromyographic activity during gait is not within the scope of this text. However, the activity of certain extrinsic and intrinsic muscles will by necessity be presented to allow a better understanding of foot control during various activities. In addition to the dynamics of the lower limb, the chapter will include a description of motion in certain joints of the 26 bones of the foot as determined by analysis of instant centers of rotation and their associated surface velocities. The necessity of understanding the function of a muscle and joint in order to effect the proper rehabilitation of the foot following injury, surgery, or immobilization is also illustrated. Examples of certain disease entities are presented to illustrate the difference between the function of the normal foot and that of a diseased or deformed foot. In Western society the foot is more often than not clothed in a semirigid covering, the shoe, and certain disease conditions develop simply because of this circumstance. When pathomechanics of such externally restricting materials are understood, a rationale for treatment of foot disorders such as bunions can be appreciated.

LENGTH OF NORMAL FOOT

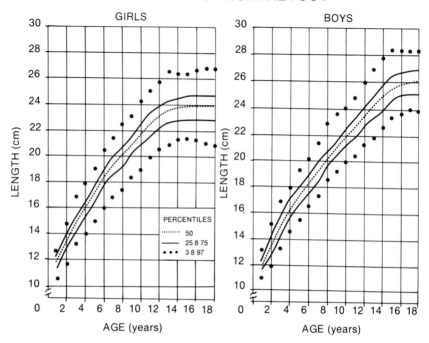

Fig. 7–1. Length of the normal foot of girls and boys derived from serial measures on 512 children from 1 to 18 years of age. (Adapted from Blais et al., 1956.)

GROWTH OF THE FOOT

The foot is formed at the time of the development of the limb buds during the eighth week of embryonic gestation. Following birth, the growth of the foot in both girls and boys proceeds at a steady rate, although it is recognized that the child as a whole has two areas of fast growth, namely the first 2 years of life and during puberty. Blais et al. (1956) showed that the foot appears to be closer to the adult size at all times during normal development of the child. On the average, at age 1 year in girls and 18 months in boys, the length of the foot is one-half the length of the respective adult foot (Fig. 7–1). This situation contrasts with that in the femur and the tibia, which do not attain half their mature length until 3 years later in both girls and boys. The relatively large size of the foot, then, is important for providing a broad base on which the child's body is supported, and this base may at times compensate for the child's lack of muscle strength and coordination.

MOTION IN THE LOWER EXTREMITY

The speed of normal walking is about 5.63 kilometers per hour. At that speed, a person averages 60 cycles per minute and spends 62% of each cycle in stance phase. Part of this time is a period of double support when both feet are on the ground, each foot in a different part of the stance phase. As the speed of gait increases, less time is spent in the period of double support and more time is spent in swing phase.

During normal walking, the entire lower extremity (including the pelvis, femur, and tibia) tends to rotate internally (inward) through the swing phase of gait into the initial 15% of stance phase. In the middle of stance phase and at push off, the entire lower extremity begins to reverse and rotate externally (outward) down to and including the talus so that at toe off, the end of stance phase, maximum external rotation is achieved in the lower extremity, including the foot. As this external rotation occurs, a degree of increased stability is reached along the medial (inside) aspect of the hip, knee, ankle, and foot. Since muscles are also contracting during this portion of the gait cycle, both ligaments and muscles stabilize the foot until it is lifted off the ground.

MOTION OF THE TARSAL BONES DURING GAIT

The motion of the separate joints within the foot is somewhat difficult to describe, since at times the foot functions as a single unit and at other times it is quite supple to accommodate various surfaces both during stance and during motion. Indeed, the potential of the normal foot as a prehensile limb has been well demonstrated by individuals with complete amelia of the upper extremities. These individuals dress, eat, and even write with their feet and toes (Swinyard, 1969).

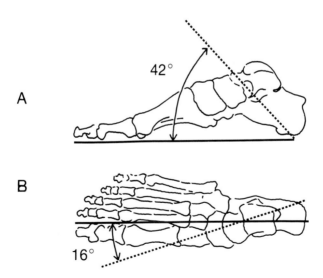

Fig. 7–2. Simplified axis of rotation of the subtalar joint.
    A. Sagittal plane (lateral view).
    B. Transverse plane (top view). (Adapted from Manter, 1941.)

THE SUBTALAR JOINT

Motion in the subtalar joint has been described by several investigators. Manter (1941), in investigating the subtalar axis of rotation, revealed that on the average the axis sits at an angle of 42 degrees with the floor from the heel upward and forward. This axis deviates 16 degrees medially from the midline of the foot (Fig. 7–2). The subtalar facets closely resemble segments of a "spiral of Archimedes," a right-handed screw in the right foot and a left-handed screw in the left foot, passing from the posterior to the anterior facets (Fig. 7–3). On the average, the subtalar joint can be inverted 20 degrees and everted about 5 degrees. Wright et al. (1964) conceived the ankle and subtalar joint complex as a universal joint functioning as a single unit. They found that throughout the stance phase of gait, the average range of motion of the subtalar joint was only 6 degrees, a so-called "functional range" of motion. The foot rests in a slightly varus position (supination) during swing phase. At heel strike, it rotates slightly into a valgus position (pronation) (Mann, 1979). The muscles of the calf and foot stabilize the subtalar joint, controlling the heel within a small range of motion.

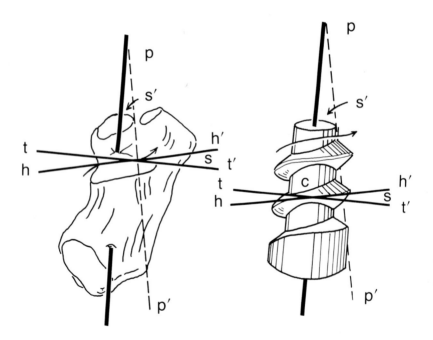

Fig. 7–3. Comparison of the posterior calcaneal facet of the right subtalar joint with a right-handed screw. Arrow represents the path of a body following the screw. hh′ is the horizontal plane in which motion is occurring. tt′ is a plane perpendicular to the axis of the screw. s is the helix angle of the screw, equal to the angle s′, which is obtained by dropping a perpendicular pp′ from the axis. (Adapted from Manter, 1941.)

THE TRANSVERSE TARSAL JOINT

The transverse tarsal joint, Chopart's joint, lies just anterior to the talus and calcaneus and is closely associated with the subtalar joint. It represents motion between the talus and navicular and the calcaneus and cuboid. Manter (1941) notes that two types of motion are achieved at this complex joint through two axes: the axis of medial-lateral rotation and the axis of flexion-extension. The former is a longitudinal axis for rotation rising up from the floor anterodorsally at an angle of 15 degrees and directed away from the midline of the foot pointing anteromedially at an angle of 9 degrees (Fig. 7–4). Around this axis internal and external rotation of the midfoot occurs during gait while the foot accommodates various plantar surfaces. Around the latter axis flexion and extension of the midfoot occur. This axis is a more oblique axis, which rises up from the floor anterodorsally about 52 degrees, and is directed away from the midline of the foot pointing anteromedially 57 degrees (Figs. 7–5 and 7–6). The axis is described for "normal" subjects and varies considerably with both the age of the individual and the shape of the foot. Therefore, this description represents at best a composite picture.

Mann and Inman (1964) have analyzed flexion and extension of the midfoot in terms of parallel axes through the talus and calcaneus (Fig. 7–7). They describe the transverse tarsal joint as functioning in a unique manner. These two axes are positioned in the frontal plane; the superior axis passes through the talar neck, and the second, inferior axis passes through the

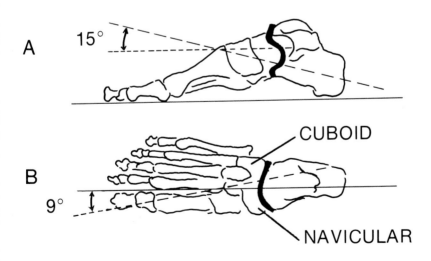

Fig. 7–4. Longitudinal axis of rotation of the transverse tarsal joint.
 A. Lateral view.
 B. Top view. (Adapted from Manter, 1941.)

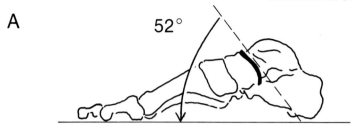

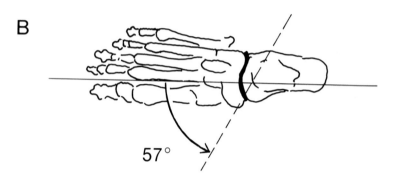

Fig. 7-5. Oblique axis of flexion and extension of the transverse tarsal joint.
A. Lateral view.
B. Top view. (Adapted from Manter, 1941.)

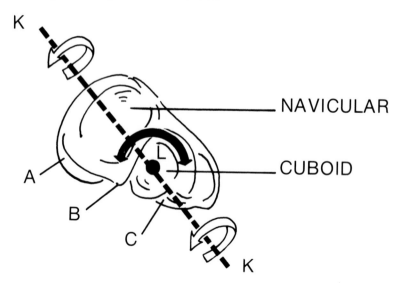

Fig. 7-6. View from posterior to anterior of the right transverse tarsal joint, showing articular surfaces of the navicular and cuboid. The longitudinal axis, L, accommodates motion as demonstrated by the solid arrows. The oblique axis, KK, provides motion as noted by the hollow arrows. A indicates the plantar calcanonavicular ligament, B indicates the deep portion of the bifurcate ligament, and C designates the long and short plantar ligaments. (Adapted from Manter, 1941.)

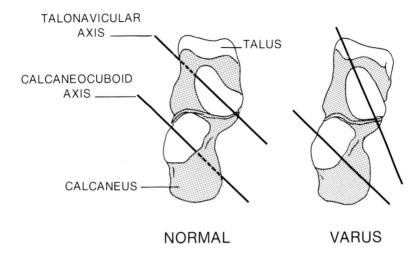

TALONAVICULAR AXIS

TALUS

CALCANEOCUBOID AXIS

CALCANEUS

NORMAL

VARUS

Fig. 7–7. Anteroposterior view of the transverse tarsal joint of the right foot. (Adapted from Mann and Inman, 1964.)

calcaneal body. These axes are concerned with motion at the talonavicular and calcaneocuboid joints, respectively. When the foot is in eversion, that is, with a tendency to be rolled flat or pronated, these two axes fall into parallel alignment. With two axes parallel in the same frontal plane, the midfoot is able to flex and extend with ease in relation to the hindfoot (heel). However, when the heel is inverted, that is, with the arch elevated or foot supinated, the axes become divergent in relation to one another. With two axes crossing one another in the same frontal plane, flexion and extension of the midfoot are significantly restricted with respect to the hindfoot. This difference in amount of flexion and extension may be one of the reasons why patients are able to tolerate a pronated foot or flatfoot more easily than they tolerate a varus or supinated foot, such as in clubfoot.

INTERTARSAL AND TARSOMETATARSAL JOINT MOTION

Motion on the surfaces of the intertarsal and tarsometatarsal joints is restricted by several factors. These include the shapes of the bones, the many restrictive ligaments, and the contracting muscles. Sliding motion occurs between the cuneiform and cuboid joints and also within the tarsometatarsal joints during gait. The axes of rotation between any two bones or group of bones within the midfoot are likely to be far from the plane of motion being examined (Fig. 7–8). Therefore, since the total excursion of any two bones of the intertarsal joints is small, for practical purposes motion may be considered as translation, or parallel motion of one surface across another.

Total motion in the midfoot ranges from just a few degrees of dorsiflexion to about 15 degrees of plantar flexion; this motion is shared by all the tarsal bones.

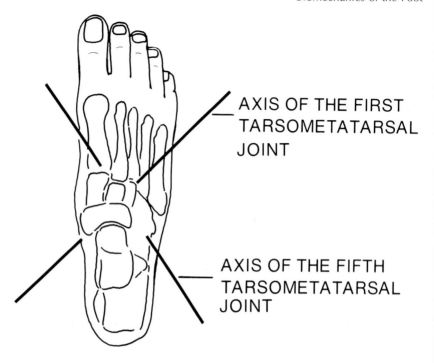

AXIS OF THE FIRST
TARSOMETATARSAL
JOINT

AXIS OF THE FIFTH
TARSOMETATARSAL
JOINT

Fig. 7–8. Tarsometatarsal axes of rotation. (Adapted from Hicks, 1953.)

The shape of the arch is affected by motion of all the tarsal joints. Of particular importance are the metatarsal-tarsal joints, Lisfranc's joint. Hicks (1953) showed that by moving the first metatarsal ray (the metatarsal and phalanges of the hallux) upward and downward, the second through the fifth lateral rays moved successively less. Conversely, when the fourth and fifth rays were moved upward and downward, the medial rays of the foot tended to move less.

The second tarsometatarsal joint is recessed into the midfoot, forming a "key-like" configuration with the middle cuneiform. This configuration restricts motions of the second ray, making it more stable when compared with the first ray and lateral three rays. The lateral rays also have a greater motion than the second ray. This relative rigidity of the second ray is important during the later stages of stance, when load is transferred into the forefoot for toe off, because it permits an increased load to be transmitted through the second metatarsal. A dramatic example resulting from this increased loading can be seen in roentgenograms of a ballet dancer's foot. Hypertrophy of the second metatarsal occurred from excessive repeated loading through the ray (Sammarco, 1980) (Fig. 7–9).

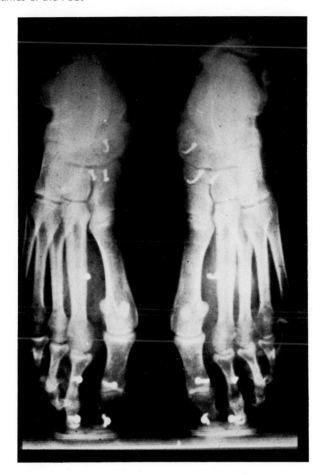

Fig. 7-9. Roentgenograms of the feet of a ballerina standing "sur les pointes." The thickened second metatarsal and the recessed second tarsometatarsal joint form a "key-like" lock which stabilizes the second ray.

THE METATARSAL BREAK

In the later part of stance phase, as the weight is transferred to the forefoot, there is an axis through which all toes extend at the metatarsophalangeal joints. This oblique axis, which overlies the metatarsophalangeal joints, is called the metatarsal break. It may vary considerably in its orientation to the long axis of the foot, from 50 to 70 degrees (Fig. 7-10), and is a generalization of the instant centers of rotation of all five metatarsophalangeal joints. The metatarsal break has been used in analyzing shoe wear and shoe fit.

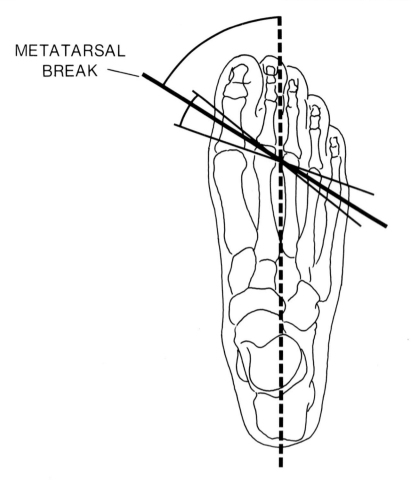

METATARSAL
BREAK

Fig. 7–10.   The metatarsal break (top view). (Adapted from Mann, 1975.)

MOTION OF THE HALLUX

Motion of the forefoot can be illustrated by the metatarsophalangeal joint motion of the hallux. The hallux must accommodate a wide range of motion for the foot to perform a great variety of tasks. The hallux metatarsophalangeal joint has a range of motion of 30 degrees of flexion to 90 degrees of extension. A hooking action with the hallux flexed to support the body, as when one stands on the edge of a step supported by the toes alone, is an example of flexion. In order to accomplish this action, the toe is flexed and all muscles must contract. It is interesting to note that modern boots used for climbing rock and timber have rough soles which "hook" onto a crag or limb while providing a stiff shank for support and a thick upper portion for protection of the foot. This design helps substitute boot function for foot function. One must also be able to extend the joint in order to accommodate a crouched position, best illustrated by a sprinter poised at the start of a race on his starting blocks. In this situation extension of the hallux metatarsophalangeal joint approaches 90 degrees. Extension of almost 90 degrees also occurs at the toe off portion of gait.

Analysis of motion of the great toe in the sagittal plane reveals that instant centers often fall within the head of the metatarsal (Fig. 7–11). Surface velocities in the sagittal plane reflected by these instant centers tend to show sliding, as in all diarthrodial joints, through most of the range of motion, that is, the joint motion occurring during most of our activity, from the few degrees of flexion through almost 90 degrees of extension. In walking the velocities are parallel to the joint surfaces, and motion occurs between the metatarsal and the proximal phalanges. As in the case of the baseball catcher crouched behind home plate, however, at extreme extension the surface velocity vectors show that the joint is being compressed or jammed. In this crouched position, a similar action is also occurring in the ankle.

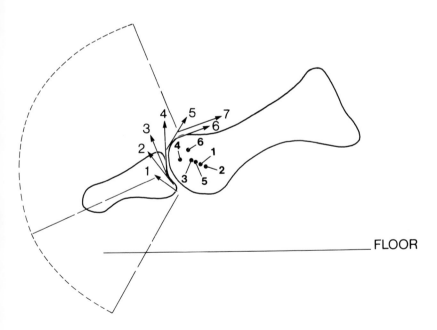

NORMAL WEIGHT BEARING

FLOOR

Fig. 7–11. Instant center and surface velocity analysis of the metatarsophalangeal joint of the hallux in the sagittal plane. Each arrow denoting surface velocity corresponds to the similarly numbered instant center. Sliding takes place throughout most of the motion except at the limit of extension, which occurs at toe off in the gait cycle and with squatting. Here compression occurs.

The collateral ligaments of the hallux metatarsophalangeal joint are somewhat loose and provide a certain degree of play, both medially and laterally. Instant centers for analysis of medial and lateral motion lie far from the joint surface (Fig. 7–12), and for practical purposes motion can be considered to be translation between the two bones.

In the development of hallux valgus, the hallux rotates internally (inwardly) and drifts into valgus angulation (points outward). The lateral and collateral ligaments contract, and the medial collateral ligaments stretch out. The metatarsal head develops a medial osteophyte, that is, a bunion. Since the hallux lies in an unnatural position, surface velocities show an altered pattern. Now jamming and distraction take place where sliding normally occurred (Figs. 7–13 and 7–14).

NORMAL WEIGHT BEARING

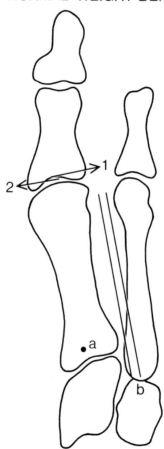

Fig. 7–12. Instant center and surface velocity analysis of the metatarsophalangeal joint of the hallux in the transverse plane. Sliding occurs at the joint surface even though the range of motion is small.

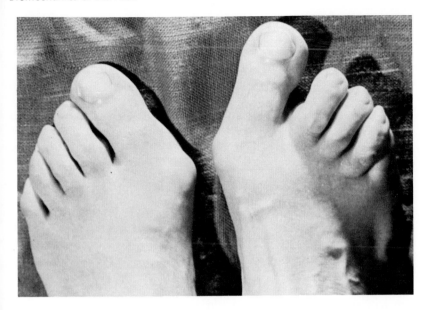

Fig. 7–13. Feet of a patient with hallux valgus and bunions. The hallux valgus is represented by the position of the great toes, which are pointed outward on both feet. Just proximal to the toes bony protuberances on each foot, the bunions, are visible.

BUNION

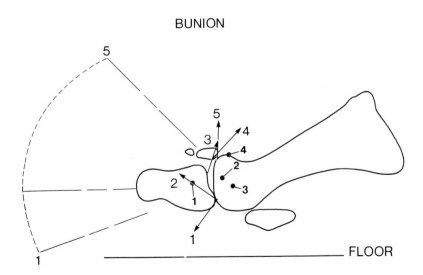

Fig. 7–14. Altered instant center and surface velocity caused by the presence of a bunion, seen through analysis of motion of the metatarsophalangeal joint in the sagittal plane.

MECHANICS OF THE SMALL TOES

The lateral four toes have three phalanges each. The length of the second toe may be less than, equal to, or greater than the length of the hallux. Motion at the metatarsophalangeal joint is approximately 90 degrees of extension and 50 degrees of flexion, or slightly greater than the hallux motion. The mechanism by which the small toes move is similar to that of the hand. Muscles which control the metatarsophalangeal and inter-phalangeal joints originate within the foot (intrinsic muscles) and in the calf (extrinsic muscles). During the latter part of stance phase, in the period up to toe off, the muscles stop functioning to allow the toes to greatly extend during push off. This extension also passively stiffens the tarsals and metatarsals by means of the plantar mechanism.

Motion of the toes in flexion is brought about at the metatarsophalangeal joint by the weak lumbricals and interossei (Sarrafian and Topouzian, 1969) (Figs. 7–15 and 7–16). But the flexor brevis and flexor digitorum longus, which insert on the middle and distal phalanges, respectively, contribute their power and provide a stronger flexion force (Fig. 7–17). The metatar-sophalangeal joints extend by the action of the extensor digitorum longus through the extensor sling which supports the proximal portion of the proximal phalanges (Fig. 7–18). The extensor digitorum communis is strong and acts as the principal motor for the action.

Extension of the middle and distal phalanges is mediated through an extensor hood (see Fig. 7–15), much the same as in the hand, and is controlled by the lumbricals and interossei.

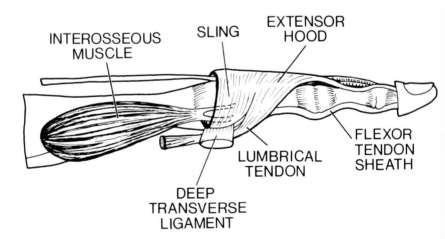

Fig. 7–15. Lateral diagram of a single toe, illustrating muscles and ligaments required for motion.

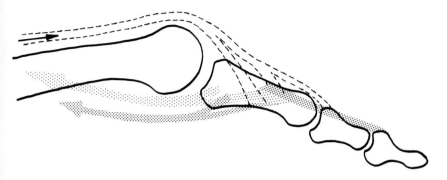

Fig. 7–16. Diagram showing the pull of the lumbricals and interossei flexing the metatar-sophalangeal joint and extending the interphalangeal joint.

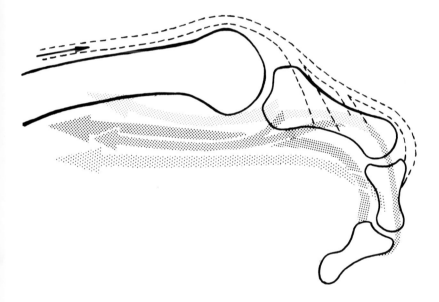

Fig. 7–17. Diagram showing the pull of the lubricals and interossei with the long and short flexors of the toe. This pull causes a clawing of the toe.

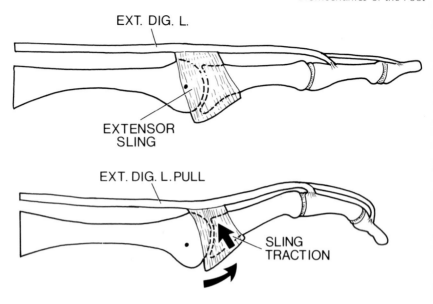

Fig. 7–18. Lateral diagram showing the action of the extensor sling. When the extensor digitorum longus contracts, the proximal phalanx is lifted through this mechanism into extension.

FUNCTION OF THE PLANTAR FASCIA

The function of the plantar fascia is also complex. From its attachment to the calcaneus the plantar fascia runs forward to span all the tarsal and metatarsophalangeal joints and to attach onto the plantar aspect of the proximal phalanges. This creates a truss-like structure whose links are the tarsal bones and ligaments of the foot, which are held at their base by a tether, that is, the plantar fascia (Fig. 7–19). Although Wright and Rennels (1964) showed that this heavy ligamentous structure will elongate slightly with increasing loads, this elongation is more of a shock-absorbing mechanism than a mechanism for obtaining motion in the foot. Since the plantar fascia spans the entire longitudinal arch and has relatively little intrinsic ability to lengthen, it acts as a cable between the heel and toes.

The subtalar, intertarsal, and tarsometatarsal joints all have small ranges of motion because of the irregular shape of each bone and the multifaceted, curved articular surfaces, which are bound by tight ligaments. The stability created ensures a good platform for standing. However, the combined motion of these joints is great because of the multitude of moving parts. There must be a mechanism for stiffening the foot during such actions as running and climbing. This mechanism is provided actively by the contraction of the calf muscles such as the triceps surae, peroneus longus, peroneus brevis, tibialis anterior, and tibialis posterior, all of which insert onto tarsal

TRUSS

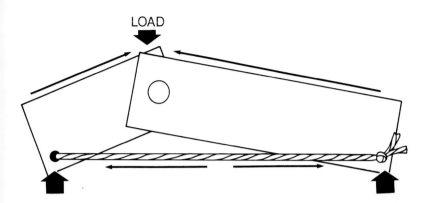

Fig. 7–19. Diagram of a truss. The wooden member is analogous to the bony structures of the foot. The plantar fascia is represented by a tether between the ends of the bone. The shorter the tether, the higher the truss is raised.

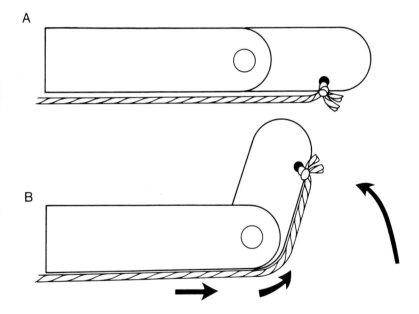

Fig. 7–20. A. Diagram of a Spanish windlass. The metatarsal is represented by the fixed wooden member, and the proximal phalanx is represented by the moving member. The rope attached to the moving member represents the attachment of the plantar fascia to the proximal phalanx.
 B. As the moving member turns, the rope advances.

bones. The intrinsic foot muscles also have their origin on the tarsal bones. When all contract, the foot becomes more rigid and acts as a stiff spring.

The mechanism by which all the tarsal joints are passively locked together is the action of the plantar fascia. A Spanish windlass mechanism is formed at the metatarsophalangeal attachment of the fascia (Fig. 7–20). As the metatarsophalangeal joints are extended passively, when one stands on the ball of the foot, the plantar fascia is pulled distally across the metatarsophalangeal joints, shortening the distance from the calcaneus to the metatarsal heads. This process makes the base of the truss shorter. As the tether is shortened and the distance between the heel and the ball of the foot is reduced, the tarsal joints are locked into a forced flexion position and the height of the longitudinal arch of the foot is increased (Fig. 7–21).

The plantar fascia also prevents the collapse of the longitudinal arch of the foot during standing because it precludes passive flexion of the toes, which would allow the fascia to relax and the arch to flatten slightly. Thus, when a welder, sprinter, or baseball catcher squats, the 12 tarsal bones of the foot are forced by this mechanism to create a solid structure to support the rest of the body.

The passive function of the plantar fascia complements the active function of the muscles, in standing, walking, running, squatting, and other activities. Surgical removal of the plantar fascia is seldom indicated and usually results in the need for an arch support for postoperative care.

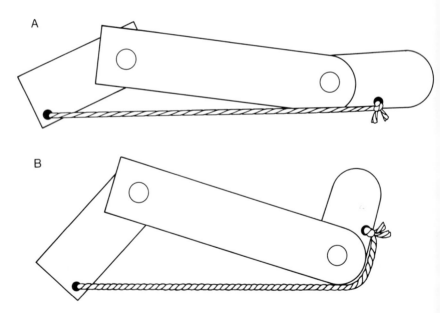

Fig. 7–21. Diagram of a combined truss and Spanish windlass **(A)**, which illustrates the function of the plantar fascia in raising the arch of the foot, at the same time locking the joints and making a single unit from multiple individual bones and joints **(B)**.

MUSCLE CONTROL IN THE FOOT

Both extrinsic and intrinsic muscles control the foot. In the calf, the gastrocnemius and soleus muscles, which combine to form the Achilles tendon inserting on the calcaneus, are the strongest flexors of the ankle. Their function is mediated through the subtalar joint, which is rather stiff in the sagittal plane and transmits the force directly to the talus and foot. The peroneus longus muscle controls, to a great extent, the downward pressure of the first metatarsal head, and it also controls the fine movements of the hallux. Alpine skiers have great need for this muscle activity to control the inner edge of their ski while turning. The peroneus brevis muscle acts as an accessory ankle flexor and stabilizes the foot laterally.

Of the muscles of the posteromedial compartment of the ankle, the tibialis posterior muscle, relatively active in normal standing, is most important in

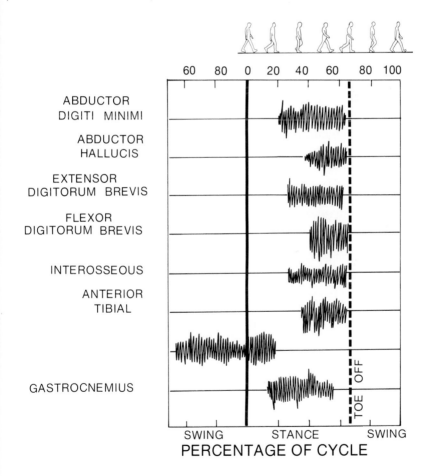

Fig. 7–22. Electromyographic tracing of the calf and foot muscles during walking. (Adapted from Mann and Inman, 1964.)

controlling the medial stability of the foot and functioning as an accessory flexor of the ankle. This muscle is of particular importance because of its special function in actively supporting the longitudinal arch. It has the function of flexing the midfoot through its pull on a broad insertion of several tarsal bones and also of supporting the talar head and neck through a sling-like action. Isolated loss of this muscle due to a severed tendon leads to collapse of the arch. The flexor hallucis longus and flexor digitorum longus control flexion of the toes during gait and are strong. Those muscles which decelerate the foot as it strikes the ground are the tibialis anterior and the extensor hallucis longus and brevis. This deceleration prevents the foot from slapping at heel strike, and thus flaccid paralysis of the muscles from disease causes a slapping-type gait (Fig. 7–22).

MUSCLE CONTROL OF THE HALLUX

Control of the great toe is mediated through the tendons of both the intrinsic and extrinsic muscles of the foot. A cross section of the proximal phalanx reveals that the relative position of the flexors, extensors, abductors, and adductors is such that these muscles move the hallux in any direction within the confines of its ligaments (Fig. 7–23). Within the tendons of the flexor hallucis brevis are the two sesamoid bones of the hallux, located directly beneath the head of the first metatarsal. The sesamoids have several functions. They act to transfer loads from the ground through the soft tissues of the forefoot to the metatarsal heads. Their position within the tendons of the flexor hallucis brevis permits them to increase the length of the moment arm, giving the muscle a greater mechanical advantage as a flexor, similar to the mechanical advantage the patella provides at the knee. The sesamoids are at a fixed distance from the proximal phalanx into which the flexor hallucis brevis inserts. Therefore, as the hallux is extended at the end of the stance phase, the sesamoids are pulled forward, and thus the correct relationship is maintained between the floor reaction forces, the soft tissue fat pads of the hallux, and the metatarsal head.

BUNIONS

Since in Western cultures shoes are usually worn, an external force is normally present along the medial side of the hallux. A normal valgus angulation of less than 15 degrees is present at the hallux metatarsophalangeal joint. However, the external force applied when a tightly fitting "style" shoe is worn over a period of years tends to increase the angle, producing hallux valgus and a bunion (see Fig. 7–13). This type of bunion differs from that caused by a disease process, such as rheumatoid arthritis. The tight tip of the "style" shoe limits hallux motion in two planes. The medial ligaments of the hallux metatarsophalangeal joint stretch out, and the

NORMAL TENDON POSITION

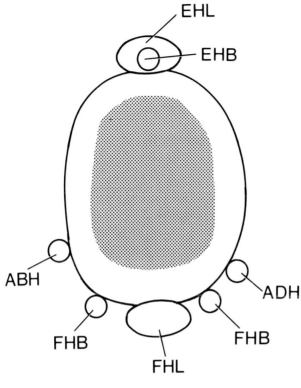

PROXIMAL PHALANX

Fig. 7–23. Diagrammatic cross section of the proximal phalanx of the hallux, showing positions of the various tendons in relationship to the bone.
EHL = extensor hallucis longus; EHB = extensor hallucis brevis; ABH = abductor hallucis; ADH = adductor hallucis; FHB = flexor hallucis brevis; FHL = flexor hallucis longus.

lateral ligaments contract. The extensor tendons drift laterally as the hallux is pushed laterally. The flexor tendons then begin to pull the first phalanx laterally as well as internally rotating it (Fig. 7–24). When this shift of the first phalanx occurs, the abductor hallucis, which normally abducts the hallux, begins to function as a flexor, causing rotation of the proximal phalanx. Balance in the toe is then lost. The first metatarsal begins to drift medially, thus forming a characteristic hallux valgus positioning. Of course, certain forms of bunions are hereditary. Reference is made here only to those forms caused by poorly fitting shoes.

NORMAL   HALLUX VALGUS

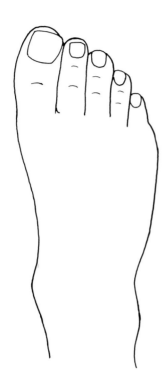
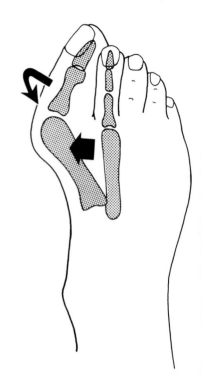

Fig. 7–24.   Diagram demonstrating the formation of hallux valgus and bunion. The pressure from the shoe forces the toe laterally. The result is a drifting of the first metatarsal medially, increasing the medial-lateral angular relation of the first metatarsal and phalanx. When this happens, the tendons which normally flex and adduct the toe begin to rotate the toes, in turn forcing the head of the first metatarsal more medially. Osteophytes then form on the medial aspect of the head of the first metatarsal.

HAMMERTOES

Shoes are a combination of design, materials, and fabrication. Despite attempts to obtain a correct fit, a shoe may be too small and the foot may be forced into a position whereby the toes are crammed together with the metatarsophalangeal joints extended and the interphalangeal joints flexed. In this situation, normal muscle balance is lost. This posturing of the toes allows the stronger calf muscles to function, causing clawing of the toes while at the same time preventing the smaller, and weaker, intrinsic muscles (lumbricals and interossei) and flexor digitorum quinti from flexing the metatarsophalangeal joints and extending the interphalangeal joints. If such positioning occurs for several years, for example, because of poorly fitted work shoes, a deformity known as hammertoe can develop with fixed

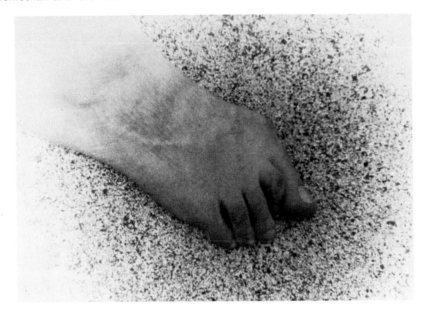

Fig. 7–25.   Hammertoe. The second toe is in extension at the metatarsophalangeal joint and in fixed flexion at the proximal interphalangeal joint.

flexion contractures of the interphalangeal joints. Dorsal dislocation at the metatarsophalangeal joint can also occur (Fig. 7–25).

LOADING CHARACTERISTICS OF THE FOOT

The distribution of loads under the foot during stance has been the subject of investigation for the last half century. Morton (1935), in his investigation on weight bearing during normal stance, showed that all metatarsal heads are in contact with the floor. This finding contrasts with those of some investigators who felt that a "transverse metatarsal arch" existed in the foot. There may indeed be a curvature of the metatarsal heads while the foot is held in the non-weight-bearing position. However, the mobility of the metatarsal heads permits them all to fall into contact with the floor as soon as load is applied during standing. In those portions of the foot having contact with the floor during normal stance, approximately 50% of the load is borne by the heel and 50% is transmitted across the heads of the metatarsals. The load on the metatarsal head of the hallux is twice that on each of the lateral four metatarsal heads (Fig. 7–26). The first metatarsal head thus transmits twice the load of each individual lateral metatarsal head, and each lateral metatarsal head bears an equal amount of the remaining portion of the load in the forefoot. A slight change in the foot structure alters the load

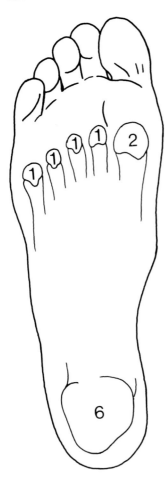

Fig. 7–26. Loads beneath the foot during normal stance. Approximately half the load is borne by the heel and the other half by the heads of the metatarsals. The load on the head of the hallux is twice that on each of the lateral four metatarsal heads.

distribution. It may also be changed with only slight modification of weight bearing, such as rocking slightly from side to side or forward and backward while standing. Thus, when standing for long periods of time, as does a West Point cadet at attention, a slight, imperceptible shifting of the weight relieves pressure on the plantar soft tissues and lessens the burning and pain of fatigue in the soft tissues of the foot.

During the later part of stance, increased loads tend to be transmitted across the second metatarsal head, as noted by Collis and Jayson (1972) and Brand (1973) (Fig. 7–27). The reason for this increase is twofold. The second metatarsal tends to be longer than the other toes; thus as heel rise occurs and the load is shifted forward, the second metatarsal head, being farther distal than the other metatarsals, tends to concentrate the gravitational force of the body. In addition, the second tarsometatarsal joint is quite stiff compared with the other midfoot joints and thus tends to yield less under loading.

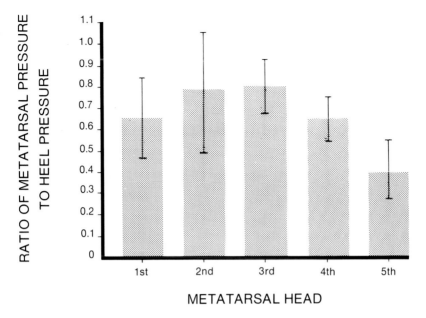

Fig. 7–27. Means and standard deviations of the ratio of metatarsal pressure to heel pressure during the later part of stance phase (5 feet). The second metatarsal heads tended to transmit the highest loads. (Adapted from Collis and Jayson, 1972.)

FOOT-FLOOR LOADS

During the stance portion of gait at heel strike, the foot-floor reaction force is located slightly lateral to the center part of the heel pad. As force is applied to the foot in the foot flat portion of stance, the progression of the force line moves laterally beneath the cuboid and the base of the fifth metatarsal. Toward the end of the stance phase, the force line courses medially beneath the second metatarsal head and then to the hallux at toe off.

Loads vary beneath the foot during sitting and standing because of the position of the foot in relation to the torso. Jones (1941) examined a subject sitting with loads applied across the knee. With all the muscles of the foot relaxed, 80% of the load was distributed to the heel, and only 20% was borne across the metatarsal heads. This difference occurred because during normal standing the load produced by the body is balanced over the center of the foot, anterior to the ankle. Slight weight shifts caused by the action of the calf muscles, the tibialis anterior and the triceps surae, adjust deviations to equilibrium.

Little activity occurs during relaxed standing, as evidenced by EMG activity (Basmajian, 1974). In the cadaver foot, where no active muscle contraction or variation of loading position occurs due to slight changes in posture, loads borne by the forefoot are equally distributed across all metatarsal heads, first through fifth. Since a small change in the loading of the foot can cause a significant change in load distribution, a small change in the muscle contraction can also cause a significant variation. The individual is thus able to combat the fatigue of standing in a single position for a prolonged period of time.

LOADS WITHIN THE FOOT

Manter (1946) analyzed the load transmitted across the individual tarsal joints during stance. He concluded that although one cannot trace an individual longitudinal pathway through which all the forces pass, the load should be greatest at the height of the longitudinal arch. Therefore, since the talonavicular and naviculocuneiform joints are the highest on the longitudinal arch, they bear the highest loads. The load is then distributed toward the heel and is transmitted to the forefoot by two general routes. The greater load is borne across the highest portion of the longitudinal arch, and it passes from the tarsal joints into the first and second metatarsals and metatarsal heads. The lesser load is transmitted somewhat laterally through the cuboid into the base of the fifth metatarsal and then to the fifth metatarsal head. These routes are described for the peak forces within the foot, however, and it should be recognized that loads are transmitted through all tarsal and metatarsal bones.

SUMMARY

1. Motion of the tarsal bones during gait is screw-like at the subtalar joint and takes place through a double axis at the transverse tarsal joint.
2. The "key-like" Lisfranc's joint in the midfoot stabilizes the second metatarsal, making it the most rigid forefoot bone and allowing it to carry most of the load during gait.
3. Instant center analysis of the hallux shows that gliding occurs throughout most of gait at the metatarsophalangeal joint with jamming at extreme extension; this surface motion is not unlike that in the ankle joint.
4. Motion in the small toes is controlled by both intrinsic and extrinsic foot muscles. The control mechanism is similar to that in the hand.
5. The plantar fascia functions through a complex truss/windlass mechanism to help passively stabilize tarsal and metatarsal bones.
6. Active control of the foot is provided by the extrinsic muscles. Anterior leg muscles decelerate the foot at heel strike, whereas posterior calf muscles propel the foot toward toe off.
7. During standing, loads are distributed equally between the heel and ball of the foot; the hallux bears twice the load of any other metatarsal. During gait, however, most of the load is transmitted through the second metatarsal.

8. During gait, foot-floor reaction forces act slightly lateral to the center of the heel at heel strike; they then shift laterally toward the cuboid, finally moving beneath the second metatarsal and hallux at toe off.

9. Loads within the foot are generally transmitted from the talus backward to the calcaneus and forward to the navicular and cuneiforms, then through the second metatarsal; only a minimal force is transmitted laterally.

REFERENCES

Basmajian, J. V.: *Muscles Alive. Their Function Revealed by Electromyography.* 3rd ed. Baltimore, Williams & Wilkins, 1974.

Blais, M. M., Green, W. T., and Anderson, M.: Lengths of the growing foot. J. Bone Joint Surg., *38A*:998, 1956.

Brand, P. W.: Personal communication, 1973.

Collis, W. J. M. F., and Jayson, M. I. V.: Measurement of pedal pressures. An illustration of a method. Ann. Rheum. Dis., *31*:215, 1972.

Hicks, J. H.: The mechanics of the foot. I. The joints. J. Anat., *87*:345, 1953.

Isman, R. E., and Inman, V. T.: Anthropometric studies of the human foot and ankle. Biomechanics Laboratory, University of California, San Francisco and Berkeley. Technical Report 58. San Francisco, The Laboratory, 1968.

Jones, R. L.: The human foot. An experimental study of its mechanics and the role of its muscles and ligaments in the support of the arch. Am. J. Anat., *68*:1, 1941.

Mann, R. A.: Biomechanics of the foot. In *Atlas of Orthotics: Biomechanical Principles and Application.* American Academy of Orthopaedic Surgeons. St. Louis, C. V. Mosby Co., 1975, pp. 257–266.

————: Personal communication, 1979.

Mann, R., and Inman, V. T.: Phasic activity of intrinsic muscles of the foot. J. Bone Joint Surg., *46A*:469, 1964.

Manter, J. T.: Distribution of compression forces in the joints of the human foot. Anat. Rec., *96*:313, 1946.

————: Movements of the subtalar and transverse tarsal joints. Anat. Rec., *80*:397, 1941.

Morton, D. J.: *The Human Foot. Its Evolution, Physiology and Functional Disorders.* New York, Columbia University Press, 1935.

Sammarco, G. J.: The foot in ballet and modern dance. In *Disorders of the Foot.* Edited by M. H. Jahss. Philadelphia, W. B. Saunders Co., 1980.

Sarrafian, S. K., and Topouzian, L. K.: Anatomy and physiology of the extensor apparatus of the toes. J. Bone Joint Surg., *51A*:669, 1969.

Swinyard, C. A.: *Limb Development and Deformity: Problems of Evaluation and Rehabilitation.* Springfield, Charles C Thomas, 1969.

Wright, D. G., and Rennels, D. C.: A study of the elastic properties of plantar fascia. J. Bone Joint Surg., *46A*:482, 1964.

Wright, D. G., DeSai, S. M., and Henderson, W. H.: Action of the subtalar and ankle-joint complex during the stance phase of walking. J. Bone Joint Surg., *46A*:361, 1964.

SUGGESTED READING

Basmajian, J. V.: *Muscles Alive. Their Function Revealed by Electromyography.* 3rd ed. Baltimore, Williams & Wilkins, 1974.

Blais, M. M., Green, W. T., and Anderson, M.: Lengths of the growing foot. J. Bone Joint Surg., *38A*:998–1000, 1956.

Collis, W. J. M. F., and Jayson, M. I. V.: Measurement of pedal pressures. An illustration of a method. Ann. Rheum. Dis., *31*:215–217, 1972.

Eyring, E. J., and Murray, W. R.: The effect of joint position on the pressure of intra-articular effusion. J. Bone Joint Surg., *46A*:1235–1241, 1964.

Frankel, V. H., and Burstein, A. H.: *Orthopaedic Biomechanics*. Philadelphia, Lea & Febiger, 1970.

Giannestras, N. J., and Sammarco, G. J.: Fractures and dislocations in the foot. In *Fractures*, Vol. 2. Edited by C. A. Rockwood and D. P. Green. Philadelphia, J. B. Lippincott Co., 1975, pp. 1400–1490.

Hicks, J. H.: The mechanics of the foot. I. The joints. J. Anat., *87*:345–357, 1953.

———: The mechanics of the foot. II. The plantar aponeurosis and the arch. J. Anat., *88*:25–31, 1954.

Hollinshead, W. H.: Knee, leg, ankle, and foot. In *Anatomy for Surgeons*. Vol. 3. Back and Limbs. New York, Harper and Row, 1969, pp. 752–874.

Isman, R. E., and Inman, V. T.: Anthropometric studies of the human foot and ankle. Biomechanics Laboratory, University of California, San Francisco and Berkeley. Technical Report 58. San Francisco, The Laboratory, 1968.

Jones, R. L.: The human foot. An experimental study of its mechanics and the role of its muscles and ligaments in the support of the arch. Am. J. Anat., *68*:1–39, 1941.

Mann, R. A.: Biomechanics of the foot. In *Atlas of Orthotics: Biomechanical Principles and Application*. American Academy of Orthopaedic Surgeons. St. Louis, C. V. Mosby Co., 1975, pp. 257–266.

Mann, R., and Inman, V. T.: Phasic activity of intrinsic muscles of the foot. J. Bone Joint Surg., *46A*:469–480, 1964.

Manter, J. T.: Distribution of compression forces in the joints of the human foot. Anat. Rec., *96*:313–321, 1946.

———: Movements of the subtalar and transverse tarsal joints. Anat. Rec., *80*:397–410, 1941.

Morton, D. J.: *The Human Foot. Its Evolution, Physiology and Functional Disorders*. New York, Columbia University Press, 1935.

Sammarco, G. J.: The foot in ballet and modern dance. In *Disorders of the Foot*. Edited by M. H. Jahass. Philadelphia, W. B. Saunders Co., 1980.

Sarrafian, S. K., and Topouzian, L. K.: Anatomy and physiology of the extensor apparatus of the toes. J. Bone Joint Surg., *51A*:669–679, 1969.

Shephard, E.: Tarsal movements. J. Bone Joint Surg., *33B*:258–263, 1951.

Swinyard, C. A.: *Limb Development and Deformity: Problems of Evaluation and Rehabilitation*. Springfield, Charles C Thomas, 1969.

Wright, D. G., and Rennels, D. C.: A study of the elastic properties of plantar fascia. J. Bone Joint Surg., *46A*:482–492, 1964.

Wright, D. G., DeSai, S. M., and Henderson, W. H.: Action of the subtalar and ankle-joint complex during the stance phase of walking. J. Bone Joint Surg., *46A*:361–382, 1964.

CHAPTER 8

Biomechanics of the Shoulder

Frederick A. Matsen III

The shoulder is defined in a broad sense as the group of structures connecting the arm to the thorax (Fig. 8–1). Its function is to position or move the humerus in space. Component parts of the shoulder are listed in Table 8–1. If this list is compared with analogous ones for other articulations, it can be seen that the shoulder is by far the most complex joint in the body. The result of the complex arrangement of structures in the shoulder is a range of motion which easily exceeds that of any other joint: the humerus can be moved through a space exceeding a hemisphere. Because of the large range of motion of the shoulder, its large number of components, and a wide variability in the size and shape of these components among individu-

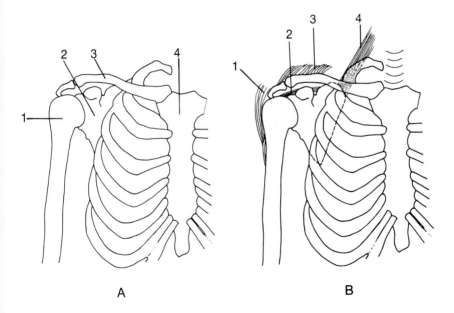

A   B

Fig. 8–1. The shoulder.
   A. Bony structures (front view): 1. humerus; 2. scapula; 3. clavicle; 4. sternum.
   B. Bony structures and some of the important muscles (front view): 1. deltoid; 2. supraspinatus; 3. trapezius; 4. levator scapulae.

TABLE 8–1 COMPONENT PARTS OF THE SHOULDER

Bones	Joints	Ligaments	Muscles
Scapula	Synovial	Glenohumeral (capsular)	Scapulohumeral and claviculohumeral
Clavicle	Glenohumeral	Coracohumeral	Superficial group
Humerus	Acromioclavicular	Coracoacromial	Deltoid
	Sternoclavicular	Acromioclavicular (capsular)	Clavicular head of pectoralis major
	Special bone-muscle-bone	Coracoclavicular	Deep group
	articulation between	Costoclavicular	Rotator cuff muscles
	scapula and thoracic	Sternoclavicular (capsular)	Subscapularis
	wall	Interclavicular	Supraspinatus
			Infraspinatus
			Teres minor
			Other
			Teres major
			Scapuloradial
			Biceps
			Scapuloulnar
			Triceps (long head)
			Thoracohumeral
			Latissimus dorsi
			Pectoralis major (sternal and
			costal heads)
			Thoracoscapular
			Serratus anterior
			Pectoralis minor
			Trapezius
			Levator scapulae
			Rhomboids
			Thoracoclavicular
			Subclavius

als, a complete, quantitative biomechanical formulation for the shoulder joint becomes impractical. However, some estimates for mechanisms and forces have been generated by greatly simplifying the actual situation.

KINEMATICS

The discussion of shoulder kinematics includes sections on (1) the total range of motion and (2) how this range of motion is achieved.

Range of Motion

The description of shoulder motion requires the definition of some special terms, such as elevation, flexion, and rotation. The published literature provides some estimates of the ranges of these motions.

Shoulder elevation is defined as the movement of the humerus away from the side in any plane and is quantitated in degrees from the vertical. Several special cases of elevation are commonly discussed: forward flexion (forward elevation in the sagittal plane) (Fig. 8–2A), abduction (elevation in the frontal plane) (Fig. 8–2B), and elevation in the "plane of the scapula" (Fig. 8–2C). Maximum elevation has been measured in males by Freedman and Munro (1966) and in females by Doody et al. (1970). Their results may be tabulated according to the percentage of normal subjects able to attain a given value of elevation (Table 8–2). These results indicate that females have a somewhat greater range than do males.

Backward elevation or extension in the posterior sagittal plane is possible to approximately 60 degrees (DePalma, 1973) (Fig. 8–3A). Depression of the arm or adduction is the action of bringing the arm to the side and is normally limited by contact with the body (Fig. 8–3B). Horizontal flexion (Fig. 8–3C) is defined as the movement of the arm forward in a horizontal plane and has a normal range of approximately 135 degrees, while movement in the opposite direction, horizontal extension, has a normal range of approximately 45 degrees (DePalma, 1973). Rotation around the axis of the humerus may be internal or external. The range of rotation varies with the position of the arm. At 90 degrees of elevation in the frontal plane, internal and external rotation may each be accomplished to about 90 degrees, yielding a maximum total range of rotation of 180 degrees (Fig. 8–3D).

How Motion is Achieved

Two features unique to the shoulder joint contribute to its large range of motion:

1. Most of the socket of the glenohumeral joint is flexible, allowing greater freedom of motion than does a rigid ball and socket joint, such as the hip.
2. Each of four articulations (glenohumeral, acromioclavicular, ster-

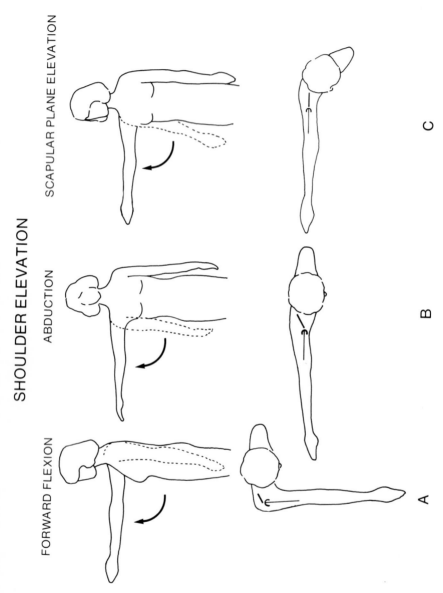

SHOULDER ELEVATION

FORWARD FLEXION   ABDUCTION   SCAPULAR PLANE ELEVATION

A   B   C

Fig. 8–2. Shoulder elevation.
A. Forward flexion. Humerus is in the sagittal plane.
B. Abduction. Humerus is in the frontal plane.

TABLE 8–2   MAXIMUM ELEVATION OF THE SHOULDER
IN THE SCAPULAR PLANE

	>180 Degrees (% of subjects)	171–180 Degrees (% of subjects)	161–170 Degrees (% of subjects)
Males*	4	33	46
Females†	28	60	12

* Data from Freedman and Munro (1966); N = 61.
† Data from Doody et al. (1970); N = 25.

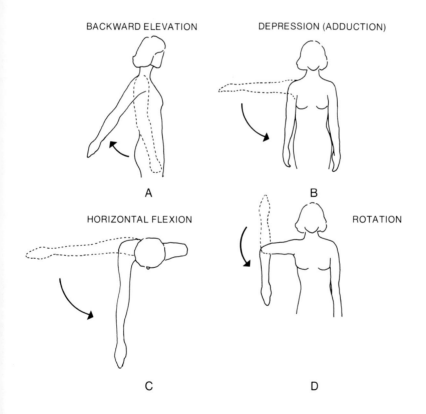

BACKWARD ELEVATION

DEPRESSION (ADDUCTION)

A

B

HORIZONTAL FLEXION

ROTATION

C

D

Fig. 8–3. A. Backward elevation.
   B. Depression (may take place in any plane).
   C. Horizontal flexion.
   D. Rotation (around the longitudinal axis of the humerus; humerus may be in any orientation).

noclavicular, and scapulothoracic) add to shoulder motion, the sum
being greater than the motion available in any single joint.

Movement of the spine can further extend the positions available to the
humerus.

The four shoulder articulations provide the motion required for full
function, such as that needed during throwing of a javelin or performance of
gymnastics on the rings. This full shoulder function involves considerably
more motion than that required for many daily activities. If the cervical
spine, forearm, wrist, and hands are normal, one can feed oneself while
keeping the whole shoulder complex immobile with the humerus at the side.
Patients with an arthrodesis of the glenohumeral joint can reach their face,
mouth, and back (Rowe, 1974). Patients with a screw through the clavicle
and coracoid process have a functional range of shoulder motion approximating 160 degrees of elevation (Rockwood and Green, 1975). Thus,
although normal motion at all four shoulder joints is required for full
function of the shoulder, considerable compensatory motion in other joints
may occur when function at one or more of the shoulder joints is limited.

Glenohumeral Joint

The glenohumeral joint consists of a nearly hemispherical convex humeral articular surface and a bony and soft tissue socket. The humeral
articular surface of the adult has a radius of curvature of from 35 to 55 mm
(Maki and Gruen, 1973). This joint surface makes an angle of 135 degrees
with the shaft and is retroverted about 30 degrees with respect to the axis of
flexion of the elbow. The socket consists of a small, pear-shaped, cartilage-
covered bony component, the glenoid fossa of the scapula, which measures
about 41 by 25 mm (Maki and Gruen, 1973). The surface area of the glenoid
is one-third to one-fourth that of the humeral head (Kent, 1971). The vertical
diameter is 75% of that of the humeral head, while the transverse diameter is
approximately 60% of that of the humeral head (Saha, 1973). Saha (1973)
found that in normal subjects 75% had a backward-facing (retrotilted)
glenoid face (average 7.4 degrees). This glenoid is lined at its perimeter by
the fibrous reflections of the capsule, the tendon of the long head of the
biceps, and the glenohumeral ligaments known collectively as the glenoid
labrum.

The capsule of the glenohumeral joint is reinforced anteriorly by the three
glenohumeral ligaments (which may appear only as capsular thickenings)
and superiorly by the coracohumeral ligament, which runs from the base of
the coracoid to the proximal end of the bicipital groove (Basmajian and
Bazant, 1959). The tendons of four muscles blend with the glenohumeral
joint capsule to form the rotator cuff. These muscles include the subscapularis, supraspinatus, infraspinatus, and teres minor.

It can be seen from the foregoing anatomical description that the socket of
the glenohumeral joint is flexible and as such allows for some freedom of

movement of the humeral head on the surface of the glenoid. This flexible socket with its small bony component increases the range of glenohumeral motion by minimizing the possibility of bone-to-bone contact.

Although the flexibility of the socket provides increased motion, it makes the glenohumeral joint more susceptible to problems of instability. Dislocations and recurrent subluxations of this joint are common. Instability may occur in an anterior, posterior, or inferior direction. Superiorly, the joint is buttressed by the coracoacromial ligament and the acromion. This arch prevents excessive upward displacement of the humeral head (Basmajian, 1969). The subacromial bursa facilitates smooth passage of the humeral head with its covering rotator cuff beneath this arch.

Joint Stability. The stability of the glenohumeral joint depends on several factors:

1. A glenoid of an adequate size. Saha (1971) found that if the ratios of the diameter of the glenoid to the diameter of the humeral head were less than 0.75 in the vertical direction and less than 0.57 in the transverse direction, the glenoid was relatively hypoplastic and the joint likely to be unstable.

2. A posteriorly tilted glenoid fossa. Saha (1971) found that 80% of 21 unstable shoulders had anteriorly tilting glenoid fossae, while the incidence of this finding in normal shoulders was 27%.

3. A posteriorly tilting humeral head (Saha, 1971).

4. An intact capsule and glenoid labrum. Reeves (1968) found that young patients with anterior shoulder instability were likely to have a detached labrum, while older patients with this condition were likely to have a stretched capsule.

5. Functional muscles controlling the anteroposterior position of the humeral head (the subscapularis and infraspinatus) (Saha, 1971).

Surface Joint Motion. In a ball-and-socket joint three types of surface motion may occur in any given plane. In the first type of motion, rotation, the socket contact point remains constant while the contact point of the ball changes as the ball rotates in the socket (analogous to the rear tire of a stuck automobile spinning in the snow) (Fig. 8–4A). In the second type, rolling, the contact point on each of the joint surfaces changes by an equal amount (analogous to an automobile tire rolling along with perfect traction) (Fig. 8–4B). In the third type, translation, the contact point on the ball remains the same while that of the socket changes (analogous to the tire of an automobile with locked brakes skidding down an icy street) (Fig. 8–4C).

The flexible glenohumeral socket with the relatively planar bony glenoid fossa allows, in addition to rotation, a small amount of translation and/or rolling, wherein the center of rotation of the humerus may displace with respect to the glenoid. Poppen and Walker (1976) measured the instant centers of rotation for elevation in the plane of the scapula and found that they lay within an average of 6 ± 2 mm of the center of the humeral head. From 0 to 30 degrees, and often from 30 to 60 degrees, the humeral ball

A. ROTATION

B. ROLLING

C. TRANSLATION

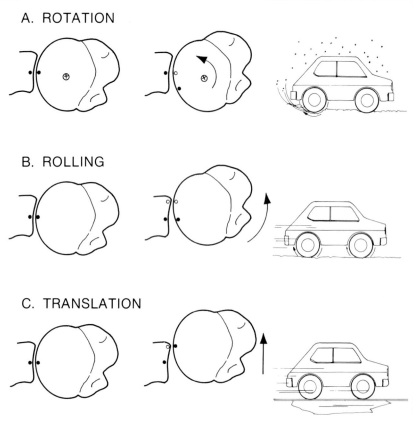

Fig. 8–4. The glenohumeral joint may allow three types of motion: rotation, rolling, and translation.

   A. In rotation the glenoid contact point remains constant while the contact point of the humeral head changes as it rotates in the glenoid (analogous to the rear tire of a stuck automobile spinning in the snow). Original contact points are indicated by the solid circles; new contact point is indicated by the hollow circle.

   B. In rolling the contact point on each of the joint surfaces changes by an equal amount (analogous to an automobile tire rolling along with perfect traction).

   C. In translation the contact point on the humeral head remains the same while that of the glenoid changes (analogous to the tire of an automobile with locked brakes skidding on ice).

moved upward with respect to the glenoid fossa by approximately 3 mm, indicating that rolling and/or translation had taken place. With each 30 degrees of further elevation, the center of the humeral head changed only 1 ± 0.5 mm up or down, indicating almost pure rotation.

Acromioclavicular Joint

   The acromioclavicular joint is a small synovial articulation between the distal clavicle and the proximal acromion (Fig. 8–5). The stability of the joint

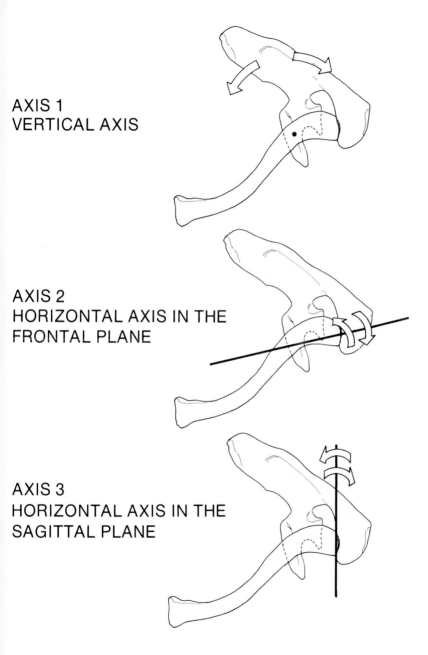

AXIS 1
VERTICAL AXIS

AXIS 2
HORIZONTAL AXIS IN THE
FRONTAL PLANE

AXIS 3
HORIZONTAL AXIS IN THE
SAGITTAL PLANE

Fig. 8–5. Axes of motion at the acromioclavicular joint (top view). Axis 1. Vertical axis for scapular rotation (through conoid ligament). Axis 2. Horizontal axis in the frontal plane for scapular rotation (through trapezoid ligament). Axis 3. Horizontal axis in the sagittal plane for scapular motion (through acromioclavicular joint).

is derived principally from the coracoclavicular ligaments. These ligaments permit scapular motion about three axes:

1. The conoid ligament runs from the knuckle of the coracoid process to the apex of the posterior arc of the distal clavicle and serves as a vertical axis for scapular rotation (Axis 1, Fig. 8–5).

2. The trapezoid ligament has linear attachments both to the upper surface of the coracoid process and to the clavicle, extending from the conoid tubercle to the acromioclavicular joint. This ligament acts as a hinge for scapular motion about a horizontal axis approximately in the frontal plane (Axis 2, Fig. 8–5).

3. The scapuloclavicular motion also takes place about a horizontal axis, passing through the acromioclavicular joint in the sagittal plane (Axis 3, Fig. 8–5) (Inman et al., 1944). It is proposed that this motion is accomplished through the relative lengthening of the posteriorly located conoid ligament when the scapula rotates posteriorly relative to the clavicle (around Axis 2).

Using a cadaver preparation consisting of only the clavicle and scapula, Dempster (1965) demonstrated a 30-degree range of rotation about the conoid ligament, a 60-degree arc for hinging on the trapezoid ligament, and a 30-degree arc for hinging about the horizontal sagittal axis through the acromioclavicular joint. The ranges of motion seen in the cadaver preparation are obviously much larger than those in living subjects, in whom the claviculoscapular motion is restricted by the thorax and muscular attachments. Inman et al. (1944) stated that the total range of acromioclavicular abduction is 20 degrees and that it occurs primarily in the first 30 and the last 45 degrees of abduction of the arm. The acromioclavicular joint usually has a meniscus, which divides it into functional units. Rotation around the conoid ligament takes place between the acromion and meniscus, and hinging on the trapezoid ligament takes place between the meniscus and the clavicle (Last, 1972).

Sternoclavicular Joint

The sternoclavicular joint is the synovial joint between the manubrium sterni and the proximal end of the clavicle. It is usually divided into two cavities by a meniscus attaching superiorly to the clavicle and inferiorly to the first costal cartilage as the cartilage attaches to the sternum. This meniscus divides the sternoclavicular joint into functional units; superior and inferior gliding occur between the clavicle and the meniscus, while anteroposterior gliding occurs between the meniscus and the sternum (Dempster, 1965).

The principal stabilizing structure of the sternoclavicular joint, the costoclavicular ligament, consists of two laminae, one running superior-medially and one running superior-laterally. These laminae secure the clavicle to the first rib (Dempster, 1965; Last, 1972). The costoclavicular ligament controls the motion between the relatively flat surfaces of this joint.

Motions allowed at the sternoclavicular joint include clavicular protraction and retraction, elevation and depression, and rotation about the longitudinal axis of the clavicle (Fig. 8–6). The fulcrum for the first two of these motions appears to be the costoclavicular ligament. Last (1972) found that when the lateral clavicle was elevated, a reciprocal depression of the medial clavicle took place, and when the lateral end of the clavicle was protracted, the sternal end moved backwards. Inman et al. (1944) observed 4 degrees of clavicular elevation for each 10 degrees of arm elevation

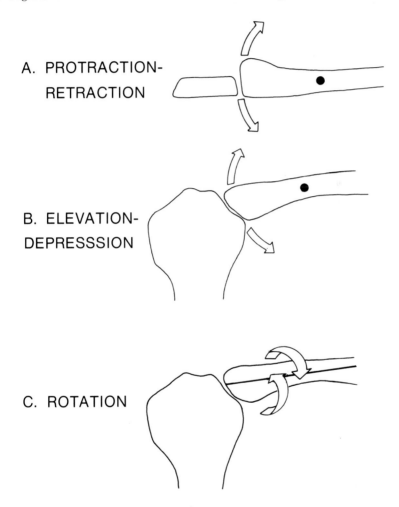

A. PROTRACTION-
   RETRACTION

B. ELEVATION-
   DEPRESSSION

C. ROTATION

Fig. 8–6. Motion at the sternoclavicular joint.
 A. Clavicular protraction and retraction in the transverse plane around a vertical axis (through costoclavicular ligament) (top view).
 B. Clavicular elevation and depression in the frontal plane around a sagittal axis (through costoclavicular ligament) (front view).
 C. Clavicular rotation around the longitudinal axis of the clavicle (front view).

through the first 90 degrees. In the second 90 degrees of arm elevation, they observed relatively little sternoclavicular motion. Rotation about the long axis of the clavicle has been found to be approximately 40 degrees (Kent, 1971; Inman et al., 1944; Last, 1972). Since much of the shoulder girdle is palpable in the normal subject, readers can confirm most of these motions by feeling their own shoulders.

Scapulothoracic Articulation

Aside from its attachment through the acromioclavicular and sternoclavicular joints, the scapula is without bony or ligamentous connection to the thorax. Its broad anterior surface is separated from the chest wall by two muscles which glide on each other during the various motions of the scapula: the subscapularis and the serratus anterior. Held in close apposition to the chest wall by its muscular attachment, the scapula may be protracted or retracted, elevated or depressed, or rotated about a variable axis perpendicular to its flat surface.

Several investigators have attempted to relate glenohumeral and scapulothoracic motion during elevation in various planes (Inman et al., 1944; Doody et al., 1970; Freedman and Munro, 1966; Poppen and Walker, 1976). Because of the infinite number of possible planes of elevation, this is obviously a complex problem; an exact solution cannot be expected for a given individual, much less for the general population with varying chest wall and scapular shapes and sizes, muscle strengths, and degrees of training. Most investigators agree on the rough estimate that if the humerus is elevated 180 degrees with respect to the thorax, approximately two-thirds of the motion of elevation takes place at the glenohumeral joint and one-third between the scapula and thorax. Inman et al. (1944) estimated that of the 60 degrees of scapular elevation, 20 degrees takes place at the acromioclavicular joint and 40 degrees takes place at the sternoclavicular joint.

Individual variability is also manifested in the initial position of the scapula. Freedman and Munro (1966) found that the base of the glenoid points down an average of 5 degrees with the arm in 0 degrees of elevation. However, the individual range was from 22 degrees down to 12 degrees up with respect to the vertical. Poppen and Walker (1976) attained a similar result. They also demonstrated that the ratio of glenohumeral to scapulothoracic motion may change with different amounts of arm elevation, glenohumeral motion being relatively more prominent during the earlier phases of elevation. Doody et al. (1970) noted that the ratio of glenohumeral to scapulothoracic motion changed when the arm was elevated with a weight in the hand. This added load tended to increase the early contribution of scapulothoracic motion.

In summary, the ratio of glenohumeral to scapulothoracic motion may vary with (1) the plane of elevation, (2) the arc of elevation, (3) the amount of load on the arm, and (4) the anatomical variations among individuals.

Fig. 8–7. The range of overhead reach is extended when the spine is tilted away from the reaching shoulder and the ribs are elevated on the reaching side.

Spine

Motion of the spine is usually not considered in a discussion of shoulder motion. It is probably worthwhile to mention, however, that the spine may play a significant role in the orientation of the arm with respect to the body's center of mass. A familiar example is the extension of the range of overhead reach when the spine is tilted away from the reaching shoulder and the ribs are elevated on the reaching side (Fig. 8–7). Atwater (1977) demonstrated the importance of spinal motion during the act of throwing. She observed that at the instant of release (instant of contact in the case of racket sports) the arm makes an angle of approximately 90 degrees with the trunk. Whether this is an "overhead" or "side arm" motion is determined by the position of the trunk and not by the position of the glenohumeral or scapulothoracic joints.

KINETICS

Although many muscles operate around the shoulder, making exact calculations difficult, it is estimated that loads at the glenohumeral joint approach body weight.

Muscle Actions

As might be surmised from the foregoing discussion of shoulder kinematics, the study of muscle actions about the shoulder has three unusual
aspects:

1. Because the glenohumeral joint lacks rigid stability, a muscle exerting
an effect on the humerus must act in concert with others to avoid producing
a dislocating force on the joint. (Compare the elbow joint, which can be
stably extended by the triceps without other muscle contraction.)

2. The existence of multiple linkages (clavicle, scapula, and humerus)
gives rise to the interesting situation in which a single muscle may span
several joints, exerting an effect on each of them. For example, the latissimus
dorsi, which originates from the chest wall and attaches to the humerus,
spans the scapulothoracic, sternoclavicular, acromioclavicular, and
glenohumeral joints.

3. The shoulder has such a large range of motion that some muscles may
have different functions depending on the initial position of the bones. For
example, the long head of the biceps may act as an accessory shoulder
abductor if the glenohumeral joint is externally rotated, whereas this
function is not possible if the humerus is initially internally rotated (Basmajian and Latif, 1957).

These three factors make it difficult to assign simple functions to the
muscles about the shoulder, to calculate individual muscle loads, and thus
to resolve force diagrams for different actions. For example, in the motion of
shoulder abduction Inman et al. (1944) and DeLuca and Forrest (1973)
recorded significant electromyographic activity in the deltoid, the clavicular
portion of the pectoralis major, the supraspinatus, the infraspinatus, the
subscapularis, the upper and middle trapezius, the serratus anterior, and the
rhomboids. When the motion of abduction was performed against resistance, significant muscle activity was also recorded from the teres major
muscle.

The actions on the shoulder musculature may be inferred from the effect
of approximation of the muscle origin and insertion. For example, in the
absence of active and passive stabilizing forces on the glenohumeral joint,
contraction of the lateral fibers of the deltoid muscle will lift the humerus
along its axis, but will not produce the motion of elevation. Elevation does
not occur because the line of action of the lateral deltoid fibers is essentially
parallel to the axis of the humerus. However, when the fulcrum of the
glenohumeral joint is more rigidly fixed by an intact and functioning
capsule, coracohumeral ligament, and rotator cuff musculature, approximations of the origin and insertion of the lateral deltoid result in the motion of
elevation. For another example, contraction of the anterior fibers of the
deltoid approximates their insertion on the deltoid tubercle and their origin
on the lateral clavicle. The resulting motion may be any combination of
elevation, horizontal flexion, internal rotation, or depression, depending on

the position of the arm and the contractile state of the other synergistic and antagonistic muscles about the shoulder.

The muscles of the rotator cuff are unique. In addition to producing joint motion by approximating their origins and insertions, they are oriented so that their tendons and muscle masses may push on the head of the humerus, around which these muscles wrap (Fig. 8–8). Thus, contraction of the subscapularis compresses the humeral head against the glenoid, internally rotates the head of the humerus, and presses against the anterior humeral head, tending to push it backward. This pushing action may be one means by which the supraspinatus prevents upward subluxation of the head of the humerus during strong contraction of the deltoid on the adducted humerus and one of the ways that a strong subscapularis prevents anterior subluxation of the shoulder.

Using the triceps brachii, Inman et al. (1944) were able to determine a direct relationship between the tension developed in the muscle and its action potential amplitude. They then used this relationship to estimate the force generated by the various shoulder muscles. Figure 8–9 is a montage of their results for the action potential amplitudes for the deltoid, clavicular portion of the pectoralis major, supraspinatus, infraspinatus, teres minor, and subscapularis in the motion of forward flexion as a function of the degree of arm elevation. Although the exact force values are not determined, one can easily see that each of these muscles contributes significantly throughout the range of shoulder flexion.

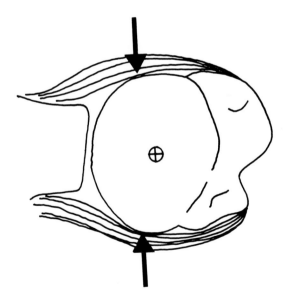

Fig. 8–8. The muscles of the rotator cuff are oriented so that their tendons and muscle masses may push on the head of the humerus, stabilizing the glenohumeral joint.

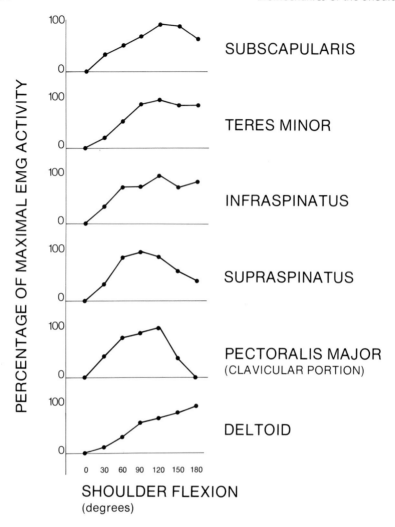

Fig. 8–9. EMG activity of various shoulder muscles during the motion of shoulder flexion, shown as a percentage of maximal EMG activity for each muscle. (Adapted from Inman et al., 1944.)

Colachis et al. (1969, 1971) used a different approach for evaluating muscle function. They measured the strength of various shoulder motions in different positions before and after suprascapular and axillary nerve blocks. Following the suprascapular nerve block (which eliminated active contraction of the supraspinatus and infraspinatus muscles), there was a 35% loss of the force of elevation at zero degrees, a 60% loss of the force of elevation at 60 degrees, and a 50% loss of the force of external rotation. An axillary nerve block reduced the force of elevation by 35% at 0 degrees and by approximately 70% at 150 degrees; external rotation strength was reduced by 45%.

Loads at the Glenohumeral Joint

Because of the large number of muscles involved in the action of elevation, and the variability in the force contributions of each of these muscles with different loads, planes of elevation, and amounts of elevation, the calculation of joint reaction forces at the glenohumeral joint becomes a formidable problem. One must therefore make simplifying assumptions to obtain estimates of the loads on this joint.

The ultimate in simplification is the static analysis of forces acting in the frontal plane proposed by Poppen and Walker (1978), in which one considers the arm at 90 degrees of elevation with only the deltoid muscle active (Fig. 8–10A). In this example, the deltoid muscle force (force M) acts parallel to the long axis of the arm. The distance between the center of rotation of the shoulder and the line of action of the muscle force (the lever arm of force M) is approximately 3 cm. The weight of the arm produces a gravitational force of 0.05 times body weight; the lever arm of this force is 30 cm. The muscle force required to keep the arm in this position (force M) is calculated to be 0.5 times body weight. Because the muscle force and the joint reaction force (force J) form a force couple, they have the same magnitude, ten times the weight of the extremity or one-half body weight.

This simplified model may provide some clinically useful information. Elevation of the arm with the elbow flexed, while not changing the mass of the extremity, would move the center of mass medially, producing a shorter lever arm for the gravitational force of 0.05 times body weight and thus reducing force M and force J by about 50% (Fig. 8–10B).

Inman et al. (1944) used a three-force system which included the force produced by the weight of the extremity, the deltoid force, and the force necessary to oppose these two combined forces. This latter force was resolved into two components: the glenohumeral joint reaction force, and the resultant rotator cuff force. They estimated that the deltoid force at 90 degrees of abduction was eight times the weight of the extremity (which they assumed to be 9% of body weight), or approximately 70% of the body weight. The joint reaction force at 90 degrees of elevation was estimated to attain a maximum of ten times the weight of the extremity, or 90% of body weight. The resultant force of the infraspinatus, teres minor, and subscapularis muscles was estimated at 9.6 times the weight of the extremity and was found to be at a maximum at 60 degrees of elevation. In a more recent study of forces in isometric abduction in the plane of the scapula, Poppen and Walker (1978) assumed that the force in a muscle was proportional to its area times the integrated electromyographic signal. Considering all the muscles to be active, they calculated that the resultant glenohumeral force reached a maximum of 90% of body weight at 90

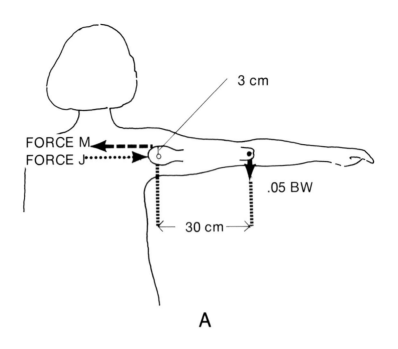

FORCE M

FORCE J

3 cm

.05 BW

30 cm

A

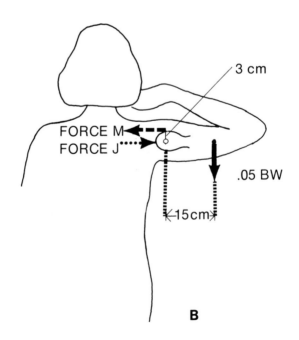

FORCE M

FORCE J

3 cm

.05 BW

15cm

B

degrees of abduction. In view of these data, it seems reasonable to consider the glenohumeral joint as a weight-bearing joint.

SUMMARY

1. The shoulder is the most mobile joint in the body because of the interposition of three linkages between the humerus and the center of mass of the body: the scapula, the clavicle, and the spine. Further shoulder motion is provided by the flexibility of the glenohumeral joint itself.
2. The stability of the glenohumeral joint depends on a glenoid of an adequate size, a posteriorly tilted glenoid fossa and humeral head, an intact capsule and glenoid labrum, and an intact and functioning rotator cuff.
3. Surface joint motion in the glenohumeral joint may include combinations of rotation, rolling, and translation.

Fig. 8–10. **A.** Estimates of the joint reaction force on the glenohumeral joint can be obtained by making simplifying assumptions (Poppen and Walker, 1978). In this example it is assumed that the arm is at 90 degrees of elevation and that only the deltoid muscle is active. The force produced through the tendon of the deltoid muscle acts at a distance of 3 cm from the center of rotation (indicated by hollow circle). The force produced by the weight of the arm is estimated to be 0.05 times body weight (BW) and acts at a distance of 30 cm from the center of rotation. The joint reaction force in the glenohumeral joint (J) may be calculated by using the equilibrium equation which states that for a body to be in equilibrium, the sum of the moments must equal zero. Moments acting clockwise are considered to be positive, while counterclockwise moments are considered to be negative.

$$\Sigma M = 0$$

$$(30 \text{ cm} \times .05 \text{ BW}) - (M \times 3 \text{ cm}) = 0$$

$$M = \frac{30 \text{ cm} \times .05 \text{ BW}}{3 \text{ cm}}.$$

Force M is approximately 0.5 times body weight. As force M and force J are almost parallel but opposite, they form a force couple and are of equal magnitude. Thus, the joint reaction force is also approximately 0.5 times body weight.

B. The arm is held in the same position, but with the elbow flexed. Elbow flexion moves the center of mass medially, shortening the lever arm of the gravitational force to 15 cm. The joint reaction force in the glenohumeral joint (J) is calculated by using the same equilibrium equation.

$$\Sigma M = 0$$

$$(15 \text{ cm} \times .05 \text{ BW}) - (M \times 3 \text{ cm}) = 0$$

$$M = \frac{15 \text{ cm} \times .05 \text{ BW}}{3 \text{ cm}}.$$

Force M is approximately 0.25 times body weight. The joint reaction force is also approximately 0.25 times body weight.

4. Although several assumptions must be made, it can be estimated that the reaction forces at the glenohumeral joint approach body weight when the unloaded arm is elevated to 90 degrees. Thus, it is apparent that the glenohumeral joint must be considered a weight-bearing joint.

REFERENCES

Atwater, A. E.: Biomechanics of throwing: Correction of common misconceptions. Paper presented at Joint Meeting of the National College Physical Education Association for Men and the National Association for Physical Education of College Women, Orlando, Florida, January 6–9, 1977.

Basmajian, J. V.: Recent advances in the functional anatomy of the upper limb. Am. J. Phys. Med., 48:165, 1969.

Basmajian, J. V., and Bazant, F. J.: Factors preventing downward dislocation of the adducted shoulder joint. An electromyographic and morphological study. J. Bone Joint Surg., 41A:1182, 1959.

Basmajian, J. V., and Latif, A.: Integrated actions and functions of the chief flexors of the elbow. A detailed electromyographic analysis. J. Bone Joint Surg., 39A:1106, 1957.

Colachis, S. C., and Strohm, B. R.: Effects of suprascapular and axillary nerve blocks on muscle force in upper extremity. Arch. Phys. Med. Rehabil., 52:22, 1971.

Colachis, S. C., Strohm, B. R., and Brechner, V. L.: Effects of axillary nerve block on muscle force in the upper extremity. Arch. Phys. Med. Rehabil., 50:647, 1969.

DeLuca, C. J., and Forrest, W. J.: Force analysis of individual muscles acting simultaneously on the shoulder joint during isometric abduction. J. Biomech., 6:385, 1973.

Dempster, W. T.: Mechanisms of shoulder movement. Arch. Phys. Med. Rehabil., 46:49, 1965.

DePalma, A. F.: *Surgery of the Shoulder.* 2nd ed. Philadelphia, J. B. Lippincott Co., 1973.

Doody, S. G., Freedman, L., and Waterland, J. C.: Shoulder movements during abduction in the scapular plane. Arch. Phys. Med. Rehabil., 51:595, 1970.

Freedman, L., and Munro, R. R.: Abduction of the arm in the scapular plane: Scapular and glenohumeral movements. A roentgenographic study. J. Bone Joint Surg., 48A:1503, 1966.

Inman, V. T., Saunders, J. B. deC. M., and Abbott, L. C.: Observations on the function of the shoulder joint. J. Bone Joint Surg., 26A:1, 1944.

Kent, B. E.: Functional anatomy of the shoulder complex. A review. Phys. Ther., 51:867, 1971.

Last, R. J.: *Anatomy. Regional and Applied.* Section 2. The Upper Limb. Edinburgh, London, and New York, Churchill Livingstone, 1972, pp. 79–111.

Maki, S., and Gruen, T.: Anthropometric study of the glenohumeral joint. Transactions of the 22nd Annual Meeting, Orthopaedic Research Society, 1:173, 1976.

Poppen, N. K., and Walker, P. S.: Normal and abnormal motion of the shoulder. J. Bone Joint Surg., 58A:195, 1976.

———: Forces at the glenohumeral joint in abduction. Clin. Orthop., 135:165, 1978.

Reeves, B.: Experiments on the tensile strength of the anterior capsular structures of the shoulder in man. J. Bone Joint Surg., 50B:858, 1968.

Rockwood, C. A.: Acromioclavicular dislocation. In *Fractures.* Vol. 1. Edited by C. A. Rockwood and D. P. Green. Philadelphia, J. B. Lippincott Co., 1975, pp. 721–756.

Rowe, C. R.: Re-evaluation of the position of the arm in arthrodesis of the shoulder in the adult. J. Bone Joint Surg., 56A:913, 1974.

Saha, A. K.: Dynamic stability of the glenohumeral joint. Acta Orthop. Scand., 42:491, 1971.

———: Mechanics of elevation of glenohumeral joint. Its application in rehabilitation of flail shoulder in upper brachial plexus injuries and poliomyelitis and in replacement of the upper humerus by prosthesis. Acta Orthop. Scand., 44:668, 1973.

SUGGESTED READING

Atwater, A. E.: Biomechanics of throwing: Correction of common misconceptions. Paper presented at Joint Meeting of the National College Physical Education Association for Men and the National Association for Physical Education of College Women, Orlando, Florida, January 6–9, 1977.

Basmajian, J. V.: Recent advances in the functional anatomy of the upper limb. Am. J. Phys. Med., *48*:165–177, 1969.

Basmajian, J. V., and Bazant, F. J.: Factors preventing downward dislocation of the adducted shoulder joint. An electromyographic and morphological study. J. Bone Joint Surg., *41A*:1182–1186, 1959.

Basmajian, J. V., and Latif, A.: Integrated actions and functions of the chief flexors of the elbow. A detailed electromyographic analysis. J. Bone Joint Surg., *39A*:1106–1118, 1957.

Bearn, J. G.: Direct observations on the function of the capsule of the sternoclavicular joint in clavicular support. J. Anat., *101*:105–170, 1967.

Broome, H. L., and Basmajian, J. V.: The function of the teres major muscle: An electromyographic study. Anat. Rec., *170*:309–310, 1971.

Chinn, C. J., Priest, J. D., and Kent, B. E.: Upper extremity range of motion, grip, strength, and girth in highly skilled tennis players. Phys. Ther., *54*:474–483, 1974.

Colachis, S. C., and Strohm, B. R.: Effects of suprascapular and axillary nerve blocks on muscle force in upper extremity. Arch. Phys. Med. Rehabil., *52*:22–29, 1971.

Colachis, S. C., Strohm, B. R., and Brechner, V. L.: Effects of axillary nerve block on muscle force in the upper extremity. Arch. Phys. Med. Rehabil., *50*:647–654, 1969.

DeLuca, C. J., and Forrest, W. J.: Force analysis of individual muscles acting simultaneously on the shoulder joint during isometric abduction. J. Biomech., *6*:385–393, 1973.

Dempster, W. T.: Mechanisms of shoulder movement. Arch. Phys. Med. Rehabil., *46*:49–70, 1965.

DePalma, A. F.: *Surgery of the Shoulder.* 2nd ed. Philadelphia, J. B. Lippincott Co., 1973.

Doody, S. G., Freedman, L., and Waterland, J. C.: Shoulder movements during abduction in the scapular plane. Arch. Phys. Med. Rehabil., *51*:595–604, 1970.

Freedman, L., and Waterland, J. C.: Shoulder movements during abduction in the scapular plane. Arch. Phys. Med. Rehabil., *51*:595–604, 1970.

Freedman, L., and Munro, R. R.: Abduction of the arm in the scapular plane: Scapular and glenohumeral movements. A roentgenographic study. J. Bone Joint Surg., *48A*:1503–1510, 1966.

Furlani, J.: Electromyographic study of the m. biceps brachii in movements at the glenohumeral joint. Acta Anat., *96*:270–284, 1976.

Inman, V. T., Saunders, J. B. deC. M., and Abbott, L. C.: Observations on the function of the shoulder joint. J. Bone Joint Surg., *26A*:1–30, 1944.

Jensen, R. K., and Bellow, D. G.: Upper extremity contraction moments and their relationship to swimming training. J. Biomech., *9*:219–225, 1976.

Jonsson, B., Olofsson, B. M., and Steffner, L. Ch.: Function of the teres major, latissimus dorsi and pectoralis major muscles. A preliminary study. Acta Morphol. Neerl. Scand., *9*:275–280, 1971/1972.

Kent, B. E.: Functional anatomy of the shoulder complex. A review. Phys. Ther., *51*:867–888, 1971.

Last, R. J.: *Anatomy. Regional and Applied.* Section 2. The Upper Limb. Edinburgh, London, and New York, Churchill Livingstone, 1972, pp. 79–111.

Lucas, D. B.: Biomechanics of the shoulder joint. Arch. Surg., *107*:425–432, 1973.

Maki, S., and Gruen, T.: Anthropometric study of the glenohumeral joint. Transactions of the 22nd Annual Meeting, Orthopaedic Research Society, *1*:173, 1976.

McMillan, J.: Therapeutic exercise for shoulder disabilities. J. Am. Phys. Ther. Assoc., *46*:1052–1067, 1966.

Moseley, H. F.: The clavicle: Its anatomy and function. Clin. Orthop., *58*:17–27, 1968.

Poppen, N. K., and Walker, P. S.: Normal and abnormal motion of the shoulder. J. Bone Joint Surg., *58A*:195–201, 1976.

———: Forces at the glenohumeral joint in abduction. Clin. Orthop., *135*:165–170, 1978.

Reeves, B.: Experiments on the tensile strength of the anterior capsular structures of the shoulder in man. J. Bone Joint Surg., *50B*:858–865, 1968.

Reid, D.: The shoulder girdle: Its function as a unit in abduction. Physiotherapy, *55*:57–59, 1969.

Rockwood, C. A.: Acromioclavicular dislocation. In *Fractures.* Vol. 1. Edited by C. A. Rockwood and D. P. Green. Philadelphia and Toronto, J. B. Lippincott Co., 1975, pp. 721–756.

Rothman, R. H., Marvel, J. P., and Heppenstall, R. B.: Anatomic considerations in the glenohumeral joint. Ortho. Clin. North Am., *6*:341–352, 1975.

Rowe, C. R.: Re-evaluation of the position of the arm in arthrodesis of the shoulder in the adult. J. Bone Joint Surg., 56A:913–922, 1974.

Saha, A. K.: Dynamic stability of the glenohumeral joint. Acta Orthop. Scand. 42:491–505, 1971.

————: Mechanics of elevation of glenohumeral joint. Its application in rehabilitation of flail shoulder in upper brachial plexus injuries and poliomyelitis and in replacement of the upper humerus by prosthesis. Acta Orthop. Scand., 44:668–678, 1973.

Shelvin, M. G., Lehmann, J. F., and Lucci, J. A.: Electromyographic study of the function of some muscles crossing the glenohumeral joint. Arch. Phys. Med. Rehabil., 50:264–270, 1969.

Singleton, M. C.: Functional anatomy of the shoulder. J. Am. Phys. Ther. Assoc., 46:1043–1051, 1966.

Walker, P. S., and Poppen, N. K.: Biomechanics of the shoulder joint during abduction in the plane of the scapula. Bull. Hosp. Joint Dis., 38:107–111, 1977.

CHAPTER 9

Biomechanics of the Elbow

Frederick A. Matsen III

The elbow is the connection of the arm to the forearm (Fig. 9–1). Its function is to position the wrist in space. The elbow is a much more stable articulation than the shoulder because it has a deeper bony socket. Component parts of the elbow are listed in Table 9–1.

KINEMATICS

The elbow allows two motions: hinging (flexion and extension) and forearm rotation (pronation and supination). This joint is essentially three articulations in one: the humeroulnar, the humeroradial, and the proximal radioulnar.

Flexion and Extension

Flexion and extension occur at the humeroulnar joint and simultaneously at the humeroradial joint. The humeroulnar joint consists of the hyperboloid-shaped humeral trochlea, which articulates with the reciprocally shaped trochlear fossa of the proximal ulna. Just lateral to the trochlea on the distal humerus is the capitellum, a spherical process which articulates with the proximal end of the head of the radius. The axis of flexion and extension passes through the middle of the trochlea, bisecting the longitudinal axis of the humerus and that of the ulna. Morrey and Chao (1976) found that the instant centers of flexion and extension varied within 2 to 3 mm of this axis.

The humeral and ulnar axes normally make an angle of 10 to 15 degrees in males and 20 to 25 degrees in females. This angle is referred to as the carrying angle. Morrey and Chao (1976) found that because the trochlea is not perfectly symmetrical, the carrying angle changes from 10 degrees of valgus to approximately 8 degrees of varus as the forearm moves from full extension to full flexion. The configuration of the humeroradial joint is actually a ball and socket, but because of the close association of this joint with the humeroulnar joint and the distal radioulnar joint, its motion is confined to two axes: flexion and extension, and pronation and supination (Steindler, 1970).

The range of flexion and extension of the elbow can be predicted from the angular values of the components, that is, the portion of an arc subtended by

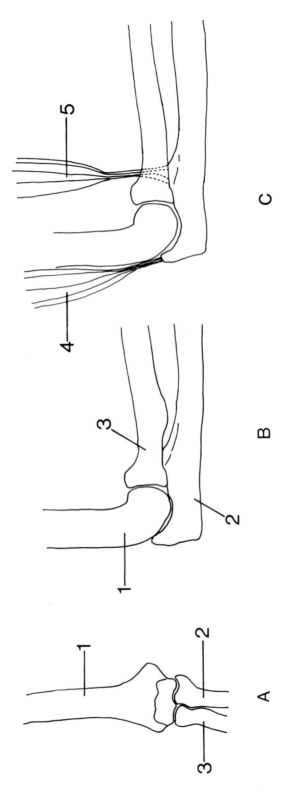

Fig. 9–1. The elbow. Bony structures; front view (**A**) and lateral view (**B**): 1. humerus; 2. ulna; 3. radius. **C.** Bony structures and some of the important muscles (lateral view): 4. triceps; 5. biceps.

TABLE 9–1 COMPONENT PARTS OF THE ELBOW

Bones	Joints	Ligaments	Muscles
Humerus	Humeroulnar	Ulnar collateral	Humeroulnar
Radius	Humeroradial	Radial collateral	Flexor
Ulna	Proximal radioulnar	Annular	Brachialis
		Interosseous	Extensors
			Triceps
			Anconeus
			Humeroradial
			Flexors
			Biceps
			Brachioradialis
			Pronator
			Pronator teres
			Supinators
			Supinator
			Biceps
			Radioulnar
			Pronators
			Pronator teres
			Pronator quadratus
			Supinator
			Supinator

the articular surfaces. The angular value of the trochlea is 330 degrees, while that of the trochlear fossa of the ulna is 190 degrees; the difference yields the range of motion of the elbow from flexion to extension: 140 degrees. (140 degrees is also the difference between the angular value for the capitulum [180 degrees] and that of the proximal radial head [40 degrees].)

Pronation and Supination

Pronation and supination are motions wherein the forearm rotates about a longitudinal axis passing through the head of the radius and the center of the distal ulna. In supination the palm is faced anteriorly if the elbow is extended, or upward if the elbow is flexed at a right angle. In pronation the palm is faced backwards or downwards. Carret et al. (1976) studied in detail the instant center of rotation at the proximal and distal radioulnar joints in different degrees of pronation and supination. They found that the proximal instant center varied with changes in the curvature of the radial head. The range of motion of the elbow in pronation and supination has been measured to be from 120 to 140 degrees (Steindler, 1970). Chao and Morrey (1978) investigated the effect of pronation and supination on the position of the ulna. They found that there was no significant axial rotation or valgus deviation of the ulna on pronation and supination of the elbow in the extended position.

Elbow Stability

Stability of the elbow results from several factors: (1) the bony interlocking of the humerus with the ulna and radius; (2) the radial (lateral) collateral ligament; (3) the annular ligament, which surrounds the neck of the radius, allowing it to rotate in pronation and supination; (4) the ulnar (medial) collateral ligament; and (5) the interosseous ligament, which connects the ulna to the radius over most of the interosseous space. While the function of the collateral ligaments is to prevent valgus and varus motion at the elbow joint, the function of the interosseous membrane is to prevent separation or longitudinal shifts of the radius and ulna with respect to each other.

KINETICS

Motions of the elbow joint include flexion, extension, pronation, and supination. In this section, the actions of the muscles responsible for these motions will be considered, as well as the resulting joint reaction forces.

Muscles and Muscle Actions

The muscles about the elbow will be discussed in terms of the four basic actions occurring at this joint—flexion, extension, pronation, and supination—although it will be seen that some muscles serve more than one action.

Flexion

The primary elbow flexors include the following muscles:

1. The brachialis, which arises from the anterior humerus and inserts on the anterior aspect of the proximal ulna.

2. The biceps, which arises via a long head tendon attached to the supraglenoid tubercle of the scapula and a short head tendon arising from the coracoid process. This muscle inserts on the bicipital tubercle of the radius, which points ulnarly in the supinated forearm and posterior-radially in the pronated forearm.

3. The brachioradialis, which originates from the lateral two-thirds of the distal humerus and inserts on the distal lateral radius near the radial styloid. It is possible that other muscles arising from the humerus and inserting on the forearm, such as the extensor carpi radialis longus and the pronator teres, serve an accessory flexor function.

There is some controversy as to which of the elbow flexor muscles are most important in different actions and in different positions of the forearm. For further details, the reader is referred to the works of Steindler (1970), Pauly et al. (1967), Basmajian (1969), and Larson (1969). Most of these studies draw their conclusions from electromyographic analysis. In interpreting them one must bear in mind the work of Currier (1972), which showed that the quantitative electromyographic method did not always correlate well with maximum muscular force.

The following qualitative statements appear to be supported by the available literature. The brachialis is a strong flexor of the forearm regardless of the amount of pronation or supination. The biceps is a strong flexor of the supinated forearm and also of the forearm in midposition; in the latter case its supinatory action is resisted by the pronator teres and pronator quadratus. The brachioradialis is a strong flexor, especially when the forearm is in midposition. Its strength is derived from its relatively long moment arm. If one attempts isometric flexion of the elbow at 90 degrees, the brachioradialis muscle may be palpated well anterior to the axis of the joint. Larson (1969) measured the isometric elbow flexor force with the elbow .flexed 65 degrees and found this force to be maximal when the forearm was supinated or in midposition, and lowest when the forearm was pronated. The respective values were 420±120 newtons, 430±120 newtons, and 390±120 newtons.

Extension

The extensors of the elbow include the three heads of the triceps and the anconeus. The long head of the triceps originates from the inferior aspect of the scapular glenoid. The other two heads arise from the posterior aspect of the humerus. The triceps inserts on the olecranon process of the ulna. The resulting triceps force lever arm significantly increases the effectiveness of the triceps in extending the elbow. The anconeus arises from the posterolateral aspect of the distal humerus and inserts on the posterolateral aspect of the proximal ulna. Pauly et al. (1967) made a detailed electromyographic study of some of the muscles crossing the elbow and concluded that the anconeus was active in initiating elbow extension, maintaining extension, and stabilizing the elbow during other motions involving the upper extremity. For example, they recorded active contraction of the anconeus during forced finger flexion and extension. Other muscles around the elbow, such as the biceps, the brachioradialis, and the triceps, also participated in this type of stabilization even though they were not required for the primary action. Apparently then, the elbow acquires additional stability by using mutually antagonistic muscles to increase the compressive load across the joint.

Little and Lehmkuhl (1966) measured the elbow extension force in 60 young females, ages 17 to 21, with the elbow in 45 degrees of flexion. The extension force ranged from 80 to 110 newtons, depending on the position of the arm. Using a cable tensometer, Currier (1972) measured the maximal isometric extension force in 41 male subjects with the elbow in different amounts of flexion. The maximal tension of 220 newtons was developed with the arm in 90 degrees of flexion.

Pronation

The pronating muscles include the pronator teres, which originates from the medial epicondyle of the humerus and attaches to the lateral aspect of the supinated radius at approximately the midposition, and the pronator quadratus, which originates from the volar aspect of the distal ulna and inserts on the lateral-distal aspect of the supinated radius. While the pronator quadratus is equally effective in different positions of elbow flexion and extension, the force produced by the pronator teres has a shorter lever arm when the elbow is completely extended. Steindler (1970) has suggested that other muscles, such as the flexor carpi radialis, may serve as accessory pronators.

Supination

Two muscles are primarily involved with supination: the supinator and the biceps. The supinator arises from the lateral epicondyle of the humerus and the lateral proximal ulna and attaches to the supinator crest on the anterior aspect of the supinated radius. The action of this muscle is relatively unaffected by the amount of flexion and extension of the elbow. The biceps, previously discussed as an elbow flexor, is considerably changed in length with elbow flexion and extension. When the biceps acts as a supinator, the action of the elbow extensors (triceps and anconeus) is required to negate its flexor action.

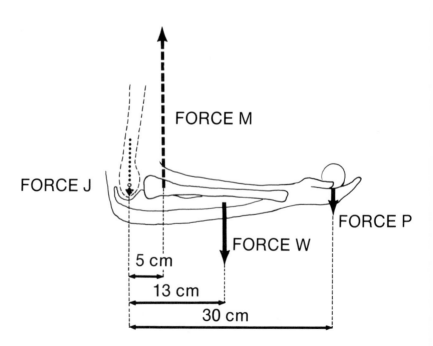

Fig. 9–2. The joint reaction force in the elbow joint during elbow flexion with and without an object in the hand can be calculated by using a simplified free body technique and the equilibrium equation which states that the sum of the moments and forces acting on the elbow joint must be zero. The primary elbow flexors are assumed to be the biceps and the brachialis muscles. The force produced through the tendons of these muscles (force M) acts at a distance of 5 cm from the center of rotation (indicated by hollow circle). The force produced by the weight of the forearm (force W), taken to be 20 newtons, acts at a distance of 13 cm from the center of rotation. The force produced by any weight held in the hand (force P) acts at a distance of 30 cm from the center of rotation.

Case A. No object is held in the hand. Force M is calculated using the equilibrium equation for moments. Clockwise moments are considered to be positive, while counterclockwise moments are considered to be negative.

$$\Sigma M = 0$$

$$(13 \text{ cm} \times W) + (30 \text{ cm} \times P) - (5 \text{ cm} \times M) = 0.$$

$$\text{If } P = 0 \text{ and } W = 20 \text{ N,}$$

$$M = \frac{13 \text{ cm} \times 20 \text{ N}}{5 \text{ cm}}.$$

Force M is calculated to be approximately 50 newtons. Force J can now be calculated by using the equilibrium equation for forces. Gravitational forces are negative, while forces in the opposite direction are positive.

$$\Sigma M = 0$$

$$M - J - W - P = 0$$

$$J = 50 \text{ N} - 20 \text{ N} - 0 \text{ N.}$$

Force J is found to be 30 newtons.

Case B. An object of 1 kg is held in the hand, producing a force of ten newtons (force P).

$$\Sigma M = 0.$$

$$\text{If } P = 10 \text{ N and } W = 20 \text{ N,}$$

$$(13 \text{ cm} \times 20 \text{ N}) + (30 \text{ cm} \times 10 \text{ N}) - (5 \text{ cm} \times M) = 0$$

$$M = \frac{260 \text{ Ncm} + 300 \text{ Ncm}}{5 \text{ cm}}.$$

Force M is found to be approximately 110 newtons.

$$\Sigma F = 0$$

$$M - W - P - J = 0$$

$$J = M - W - P$$

$$J = 110 \text{ N} - 20 \text{ N} - 10 \text{ N.}$$

Force J is found to be 80 newtons.

Thus, in this example a 1-kg object held in the hand with the elbow flexed to 90 degrees increases the joint reaction force by 50 newtons.

Elbow Motion Required for Various Activities

Different ranges of motion are required for various activities involving the elbow. Pushing up from a chair or walking on crutches requires nearly full extension. Eating and making up one's face require much elbow flexion. Opening doors and accepting coins require ample supination. Writing and ironing require pronation of the forearm.

Joint Forces in the Elbow

Many muscles are operant in elbow flexion and extension; yet a few simple approximations permit one to estimate the elbow joint reaction force in static and dynamic situations.

In the following static example, the three main coplanar forces acting on the elbow joint are analyzed. The elbow joint is flexed 90 degrees; it is assumed that the predominant elbow flexors are the biceps and brachialis and that the force produced through the tendons of these muscles (force M) acts perpendicular to the long axis of the forearm (Fig. 9–2, Case A). The distance between the center of rotation of the elbow and the point of insertion of the tendons of these muscles (the lever arm of force M) is approximately 5 cm. The weight of the forearm produces a gravitational force (force W), and the lever arm of force W is taken to be 13 cm. If the mass of the forearm is 2 kg, then the muscle force required to keep the elbow in this position (force M) is 50 newtons, and the joint reaction force on the trochlear fossa (force J) is 30 newtons. By contrast, when a 1-kg weight is held in the hand, producing a gravitational force (force P) of ten newtons at a distance of 30 cm from the center of rotation, the required muscle force (force M) is 110 newtons and the joint reaction force (force J) is over twice as high, 80 newtons (Fig. 9–2, Case B). Thus, small loads applied to the hand dramatically increase the joint reaction force.

An estimation can also be made for the joint reaction force in the elbow during extension (Fig. 9–3). In the following example, the elbow is held in 90 degrees of flexion and the arm is positioned over the head. The three main coplanar forces acting on the elbow joint are the force produced by the weight of the forearm, the tensile force through the tendon of the triceps muscle, and the joint reaction force on the trochlear fossa of the ulna. The distance between the center of rotation of the elbow and the point of insertion of the tendon of the triceps muscle (the lever arm of force M) is approximately 3 cm, 2 cm less than the lever arm of force M during elbow flexion. Because the lever arm of force M is shorter, the joint reaction force in this example is higher during elbow extension than during flexion, 110 newtons compared with 30 newtons.

Nicol et al. (1977) estimated the joint reaction force at the elbow during dressing and eating to be 300 newtons. Because an individual weighing 70 kg generates a gravitational force of almost 700 newtons, it can be seen that even these minimally stressful activities give rise to joint loads of approxi-

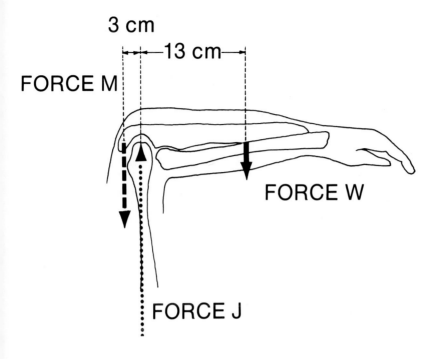

Fig. 9–3. The joint reaction force during elbow extension can be calculated by the same method:

$$\Sigma M = 0$$

$$(13 \text{ cm} \times W) - (3 \text{ cm} \times M) = 0.$$

$$\text{If } W = 20 \text{ N,}$$

$$M = \frac{13 \text{ cm} \times 20 \text{ N}}{3 \text{ cm}}.$$

Force M is found to be approximately 90 newtons.

$$\Sigma F = 0$$

$$J - M - W = 0$$

$$J = M + W$$

$$J = 90 \text{ N} + 20 \text{ N}.$$

Force J is found to be 110 newtons.

Thus, in this example the joint reaction force during elbow extension is 80 newtons higher than during elbow flexion.

mately one-half body weight. Activities such as pulling a heavy object or assisted standing from the seated position produced peak joint reaction forces of up to 1,700 newtons.

SUMMARY

1. The elbow joint is essentially three articulations in one: the humeroulnar, the humeroradial, and the proximal radioulnar.
2. This joint provides two types of motion: hinging (flexion and extension) and forearm rotation (pronation and supination).
3. Stability of the elbow joint is provided by the interlocking of the humerus, radius, and ulna and the ligamentous apparatus surrounding the joint.
4. The relatively large number of muscles producing the various motions in the elbow complicates an exact force analysis for this joint. The best estimates available suggest that static loads approach, and dynamic loads exceed, body weight.

REFERENCES

Basmajian, J. V.: Recent advances in the functional anatomy of the upper limb. Am. J. Phys. Med., 48:165, 1969.

Carret, J.-P., Fischer, L. P., Gonon, G. P., and Dimnet, J.: Etude cinematique de la prosupination au niveau des articulations radiocubitales (radio ulnaris). Bull. Assoc. Anat., 60:279, 1976.

Chao, E. Y., and Morrey, B. F.: Three-dimensional rotation of the elbow. J. Biomech., 11:57, 1978.

Currier, D. P.: Maximal isometric tension of the elbow extensors at varied positions. Part 2. Assessment of extensor components by quantitative electromyography. Phys. Ther., 52:1265, 1972.

Larson, R. F.: Forearm positioning on maximal elbow-flexor force. Phys. Ther., 49:748, 1969.

Little, A. D., and Lehmkuhl, D.: Elbow extension force. Measured in three test positions. J. Am. Phys. Ther. Assoc., 46:7, 1966.

Morrey, B. F., and Chao, E. Y. S.: Passive motion of the elbow joint. A biomechanical analysis. J. Bone Joint Surg., 58A:501, 1976.

Nicol, A. C., Berme, N., and Paul, J. P.: A biomechanical analysis of elbow joint function. In *Institution of Mechanical Engineers Conference Publications 1977–5,* Joint Replacement in the Upper Limb, pp. 45–51. Conference sponsored by the Medical Engineering Section of the Institution of Mechanical Engineers and the British Orthopaedic Association, London, April 18–20, 1977.

Pauly, J. E., Rushing, J. L., and Scheving, L. E.: An electromyographic study of some muscles crossing the elbow joint. Anat. Rec., 159:47, 1967.

Steindler, A.: *Kinesiology of the Human Body Under Normal and Pathological Conditions.* Springfield, Charles C Thomas, 1970.

SUGGESTED READING

Basmajian, J. V.: Recent advances in the functional anatomy of the upper limb. Am. J. Phys. Med., 48:165–177, 1969.

Basmajian, J. V., and Latif, M. A.: Integrated actions and functions of the chief flexors of the elbow. J. Bone Joint Surg., 39A:1106–1118, 1957.

Carret, J.-P., Fischer, L. P., Gonon, G. P., and Dimnet, J.: Etude cinematique de la prosupination au niveau des articulations radiocubitales (radio ulnaris). Bull. Assoc. Anat., 60:279–295, 1976.

Chao, E. Y., and Morrey, B. F.: Three-dimensional rotation of the elbow. J. Biomech., 11:57–73, 1978.

Currier, D. P.: Maximal isometric tension of the elbow extensors at varied positions. Part 2. Assessment of extensor components by quantitative electromyography. Phys. Ther., 52:1265–1276, 1972.

Dehaven, K. E., and Evarts, C. M.: Throwing injuries of the elbow in athletes. Orthop. Clin. North Am., 4:801–808, 1973.

Larson, R. F.: Forearm positioning on maximal elbow-flexor force. Phys. Ther., 49:748–756, 1969.

Little, A. D., and Lehmkuhl, D.: Elbow extension force. Measured in three test positions. J. Am. Phys. Ther. Assoc., 46:7–17, 1966.

Morrey, B. F., and Chao, E. Y. S.: Passive motion of the elbow joint. A biomechanical analysis. J. Bone Joint Surg., 58A:501–508, 1976.

Nicol, A. C., Berme, N., and Paul, J. P.: A biomechanical analysis of elbow joint function. In *Institution of Mechanical Engineers Conference Publications 1977–5*, Joint Replacement in the Upper Limb, pp. 45–51. Conference sponsored by the Medical Engineering Section of the Institution of Mechanical Engineers and the British Orthopaedic Association, London, April 18–20, 1977.

Pauly, J. E., Rushing, J. L., and Scheving, L. E.: An electromyographic study of some muscles crossing the elbow joint. Anat. Rec., 159:47–54, 1967.

Ray, R. D., Johnson, R. J., and Jameson, R. M.: Rotation of the forearm. An experimental study of pronation and supination. J. Bone Joint Surg., 33A:993–996, 1951.

Steindler, A.: *Kinesiology of the Human Body Under Normal and Pathological Conditions.* Springfield, Charles C Thomas, 1970.

CHAPTER 10

Biomechanics of the Lumbar Spine

Margareta Lindh

The human spine is a complex structure. Its main functions are to protect the spinal cord and to transfer loads from the head and trunk to the pelvis. The 24 vertebrae which articulate with each other permit motion in three planes. Stability is provided to the spine both intrinsically and extrinsically; ligaments and discs provide intrinsic stability and muscles give extrinsic support.

This chapter will describe the basic characteristics of the various structures of the spine and the interaction of these structures during normal spinal function. It will also discuss motion of the thoracic and lumbar spine and loads on the lumbar spine. The discussion of kinetics is limited to the lumbar spine because this area, subjected to significantly greater loads than the rest of the spine, has been given the most attention clinically and experimentally. The information presented in this chapter has been selected to give an understanding of some fundamental aspects of biomechanics of the spine which can be put into practical use in daily activities.

FUNCTIONAL UNIT OF THE SPINE: THE MOTION SEGMENT

The functional unit of the spine is the motion segment, which consists of two vertebrae and their intervening soft tissues (Fig. 10–1). Two superimposed vertebral bodies, the intervertebral disc, and the longitudinal ligaments constitute the anterior portion of this segment. The corresponding vertebral arches, the intervertebral joints, the transverse and spinous processes, and ligaments compose the posterior portion. The arches and the vertebral bodies form the vertebral canal, which protects the spinal cord.

Anterior Portion of the Motion Segment

The vertebral bodies are designed to sustain mainly compressive loads and become larger as the superimposed weight of the upper body increases. The vertebral bodies in the lumbar spine have a greater height and cross-sectional area than those in the thoracic and cervical spine; their increased size allows them to sustain the greater loads to which this part of the spine is subjected.

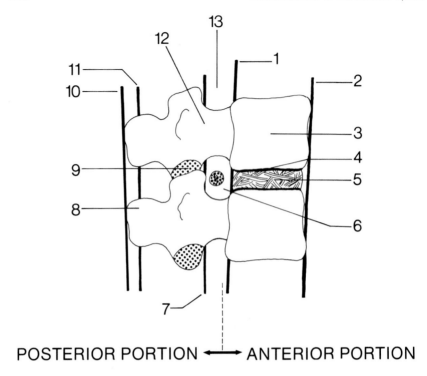

POSTERIOR PORTION ⟵⟶ ANTERIOR PORTION

Fig. 10–1. Functional unit of the spine: the motion segment. 1. Posterior longitudinal ligament; 2. anterior longitudinal ligament; 3. vertebral body; 4. cartilaginous end-plate; 5. intervertebral disc; 6. intervertebral foramen with nerve root; 7. ligamentum flavum; 8. spinous process; 9. intervertebral joint and facet (with capsular ligament); 10. supraspinous ligament; 11. interspinous ligament; 12. transverse process (with intertransverse ligament); 13. vertebral canal (with spinal cord).

The intervertebral disc is of great mechanical and functional importance. It is composed of two structures: an inner portion, the nucleus pulposus, and an outer portion, the annulus fibrosus (Fig. 10–2A). The nucleus pulposus is a fluid mass composed of a colloidal gel rich in water-binding glucosaminoglycans. It is centrally located in the disc except in the lower lumbar segments, where its location is more posterior. The annulus fibrosus consists of fibroid cartilage with bundles of collagen fibers, arranged in a criss-cross pattern (see Fig. 10–1), which allows them to withstand high bending and torsional loads. The cartilaginous end-plate, an integral part of the disc composed of hyaline cartilage, separates the nucleus pulposus and the annulus fibrosus from the vertebral body (see Fig. 10–1).

During daily activities the disc is loaded in a complex manner, usually by a combination of compression, bending, and torsion. Flexion, extension, and lateral flexion of the spine produce mainly tensile and compressive stresses in the disc, whereas rotation produces shear stress.

If an incision is made through the disc, the nucleus will protrude. This protrusion shows that the nucleus is under pressure. The disc tends to separate the vertebral bodies and causes tension in the annular fibers and the longitudinal ligaments. The normal nucleus pulposus acts hydrostatically (Nachemson, 1960), and during loading the pressure is uniformly distributed. Hence, the disc provides a hydrostatic function in the motion segment,

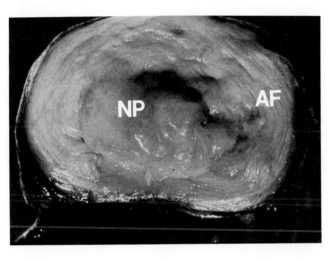

A

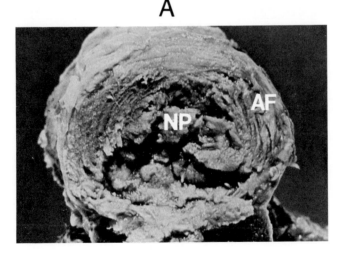

B

Fig. 10–2. Intervertebral disc composed of the nucleus pulposus (NP) and the annulus fibrosus (AF).
A. Normal disc.
B. Severely degenerated disc.

acting as a cushion between the vertebral bodies to store energy and distribute loads.

Measurement of the intradiscal pressure in normal and slightly degenerated lumbar nuclei pulposi from autopsy subjects has shown an intrinsic pressure in the unloaded disc of about 10 newtons per sq. cm (Nachemson, 1960). This prestress in the disc is due to ligament forces. The pressure in a disc loaded in compression has been shown to be about 1.5 times the externally applied load per unit area. The nuclear material has been found to be only slightly compressible. Thus, a compressive load makes the disc bulge laterally, and circumferential tension is placed upon the annular fibers. The tensile stress in the posterior part of the annulus fibrosus in the lumbar spine has been estimated to be four to five times the applied axial load (Nachemson, 1960, 1963; Galante, 1967) (Fig. 10–3). The thoracic annuli fibrosi are subjected to less tensile stress than the lumbar annuli fibrosi because of differences in geometry. The ratio of disc diameter to height is higher in the thoracic discs than in the lumbar discs; thus, the circumferential stresses become less dominant in the thoracic discs (Kulak et al., 1975).

Degeneration of the disc with age (Fig. 10–2B) reduces its water-binding capacity. As the disc becomes dryer, its elasticity decreases. It gradually loses the ability to store energy and to distribute stresses, and is therefore less capable of resisting loads.

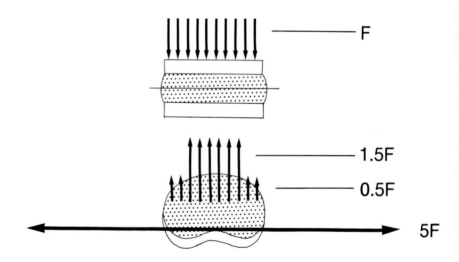

Fig. 10–3. Distribution of stress in a lumbar disc under compressive loading. The compressive stress is highest in the nucleus pulposus, 1.5 times the externally applied load (F) per unit area. In the annulus fibrosus the compressive stress is only about 0.5 times the externally applied load, and this structure is mainly stressed in tension. The tensile stress on the annulus is four to five times the externally applied load. (Adapted from Nachemson, 1975.)

Posterior Portion of the Motion Segment

The posterior portion guides the motion of the motion segment. The direction of this motion is determined by the orientation of the facets of the intervertebral joints. Throughout the spine this orientation changes in relation to the transverse and frontal planes (Fig. 10–4). Except for the two uppermost cervical vertebrae, whose facets are oriented in the transverse plane, the facets of the intervertebral joints of the cervical spine are oriented 45 degrees to the transverse plane and parallel to the frontal plane (Fig.

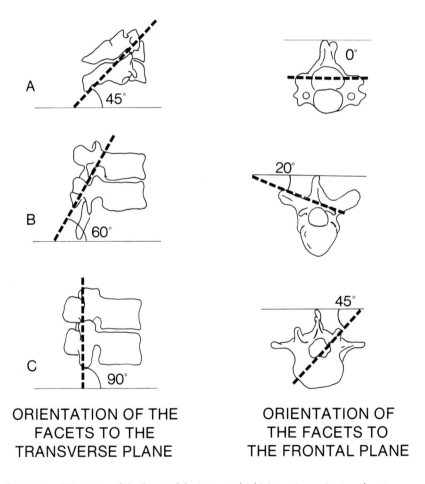

ORIENTATION OF THE
FACETS TO THE
TRANSVERSE PLANE

ORIENTATION OF
THE FACETS TO
THE FRONTAL PLANE

Fig. 10–4. Orientation of the facets of the intervertebral joints. (Approximate values.)
  A. Lower cervical spine. The facets are oriented 45 degrees to the transverse plane and parallel to the frontal plane.
  B. Thoracic spine. The facets are oriented 60 degrees to the transverse plane and 20 degrees to the frontal plane.
  C. Lumbar spine. The facets are oriented 90 degrees to the transverse plane and 45 degrees to the frontal plane.

10–4A). The alignment of these cervical intervertebral joints (C3–C7) allows flexion, extension, lateral flexion, and rotation. The facets of the thoracic joints are oriented 60 degrees to the transverse plane and 20 degrees to the frontal plane (Fig. 10–4B), allowing lateral flexion, rotation, and some flexion and extension. In the lumbar region the intervertebral facets are oriented at right angles to the transverse plane and 45 degrees to the frontal plane (Fig. 10–4C) (White and Panjabi, 1978). This alignment allows flexion, extension, and lateral flexion, but almost no rotation. The lumbosacral intervertebral joints differ from the other intervertebral joints in the lumbar area. The orientation and the shape of the facets at this level allow some rotation (Lumsden and Morris, 1968). It should be mentioned that the values given here are only approximate, as variations in facet orientation are found both within an individual and among individuals.

The facets were previously thought to be mainly involved in guiding motion of the motion segment, having only a modest load-bearing function. In recent studies, however, their load-bearing function has been shown to be more complex. Depending on the position of the spine, the load-sharing between the facets and the disc varies, with the facets bearing between zero and about 30% of the load. The load-bearing function of the facets is particularly evident when the spine is in a hyperextended position (King et al., 1975). The importance of the arches and intervertebral joints in resisting shear forces is exemplified by the risk of a forward displacement of the vertebral body in conditions of spondylolysis or defective joints.

The transverse and spinous processes serve as sites of attachment for the spinal muscles, whose activity initiates motion and provides extrinsic stability to the spine.

Ligaments of the Spine

The ligamentous apparatus provides part of the intrinsic stability of the spine. Most ligaments of the spine are composed predominantly of collagen fibers, which minimize elongation. The ligamentum flavum, which connects the vertebral arches longitudinally, is an exception with its high percentage of elastic fibers. The elasticity of the ligamentum flavum allows it to shorten during extension of the spine and to lengthen during flexion, and thus this ligamentum is under constant tension. From its distance to the center of motion in the disc, it prestresses the disc, i.e., an intradiscal pressure is created, providing intrinsic support to the spine (Rolander, 1966; Nachemson and Evans, 1968).

Additional functions of the ligaments are to transfer tensile loads from one vertebra to a another and to allow smooth motion within the physiologic range to be carried out with a minimum of resistance.

KINEMATICS

Motion in the spine is produced by the coordinated action of nerves and muscles. Agonistic muscles (prime movers) initiate and carry out motion, whereas antagonistic muscles often control and modify it. The range of motion differs at various levels of the spine, depending on the orientation of the facets of the intervertebral joints at each level. The motion between two vertebrae is small and does not occur independently. Spinal movements are always a combined action of several segments.

Skeletal structures that influence motion of the spine are the rib cage, which limits thoracic motion, and the pelvis, whose tilting increases trunk movements.

Range of Segmental Motion of the Spine

The vertebrae have 6 degrees of freedom, that is, rotation about and translation along a transverse, a sagittal, and a longitudinal axis. Although the instantaneous center of motion in the lumbar spine has been shown to lie within the disc under normal conditions (Rolander, 1966; Cossette et al., 1971); a location outside the disc in certain situations has been reported (Reichmann et al., 1972).

The range of motion in an individual segment of the spine has been found to vary in different studies using autopsy material or radiographic in vivo measurements, but there is agreement on the relative amount of motion at different levels of the spine. Representative values (White and Panjabi, 1978) are presented here to allow a comparison of the motion at the various levels of the thoracic and lumbar spine. A representative value for the flexion-extension range is 4 degrees in the upper thoracic motion segments, 6 degrees in the midthoracic region, and 12 degrees in the two lower thoracic segments. The range of flexion-extension progressively increases in the lumbar motion segments, reaching a maximum of 20 degrees at the lumbosacral level (Fig. 10–5).

Lateral flexion shows the greatest range in the lower thoracic segments, reaching 8 to 9 degrees. In the upper thoracic segments the range is uniformly 6 degrees. Six degrees of lateral flexion is also found in the lumbar segments except for the lumbosacral segment, where only 3 degrees of motion occurs (Fig. 10–5).

Rotation is greatest in the upper segments of the thoracic spine, where a range of 9 degrees is found. The range of motion progressively decreases caudally, reaching 2 degrees in the lower segments of the lumbar spine, but again increases in the lumbosacral segment to 5 degrees (Fig. 10–5). Representative values for motion in the lower cervical spine are also shown in Figure 10–5 for the sake of comparison.

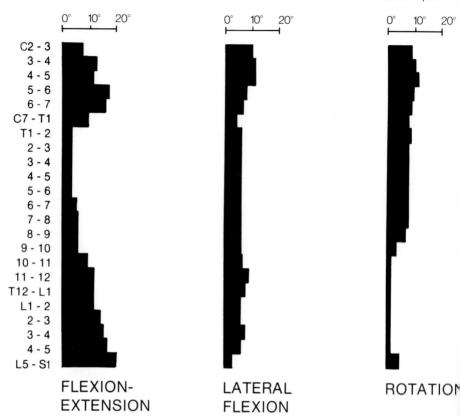

Fig. 10–5. Representative values for spinal motion are given to allow a comparison of the type and range of motion at different levels of the spine. (Adapted from White and Panjabi, 1978.)

Functional Motion of the Spine

Segmental motion cannot be measured clinically. Any motion of the spine is a combined action of several motion segments. Normal values of functional range of motion of the spine do not exist because there are great variations among individuals; in fact, the range of motion in each of the three planes shows a Gaussian distribution. The range of motion also differs between the sexes and is strongly age-dependent, decreasing by about 50% from youth to old age (Moll and Wright, 1971).

Flexion and Extension

The first 50 to 60 degrees of spinal flexion occurs in the lumbar spine, mainly in the lower motion segments (Farfan, 1975). Forward tilting of the pelvis allows further flexion (Fig. 10–6). The thoracic spine contributes little

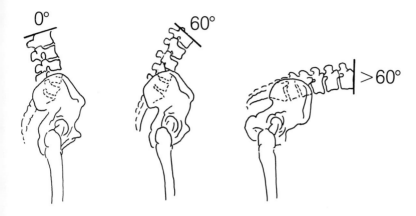

Fig. 10–6. The first 50 to 60 degrees of flexion occurs in the lumbar spine. Additional flexion of the trunk is mainly accomplished by forward tilting of the pelvis. (Adapted from Farfan, 1975.)

to flexion of the total spine because of the orientation of the facets, the almost vertical orientation of the spinous processes, and the restriction to motion imposed by the rib cage. Flexion is initiated by the abdominal muscles and the vertebral portion of the psoas muscle. The weight of the upper body then produces further flexion, which is controlled by the gradually increasing activity of the erector spinae muscles as the moment of force increases. The posterior hip muscles are active in controlling the forward tilting of the pelvis (Carlsöö, 1961). In full flexion, however, the erector spinae muscles become inactive. In this position the forward-bending moment may be passively balanced by the posterior ligaments which, initially slack, become taut at this point due to spinal elongation (Farfan, 1975).

From full flexion to the upright position of the trunk, a reverse sequence is observed. The pelvis tilts backward and the spine then extends. In some studies the concentric work performed by the muscles involved in raising the trunk has been shown to be greater than the eccentric work performed by the muscles during flexion (Friedebold, 1958; Joseph, 1960). When the trunk is extended from the upright position, the back muscles are active during the initial phase of the motion. This initial burst of activity decreases during further extension, and the abdominal muscles become active to control and modify the motion. In extreme or forced extension, activity of the extensor muscles is again required.

Lateral Flexion and Rotation

During lateral flexion of the trunk, motion may predominate in the thoracic or lumbar spine. Although the shape of the facet joints in the thoracic spine allows lateral flexion, the motion is restricted by the rib cage to a varying degree among individuals; in the lumbar spine the wedge-shaped spaces of the intervertebral joints show variations during motion (Reichmann, 1970/71). Both of these factors influence the range of motion.

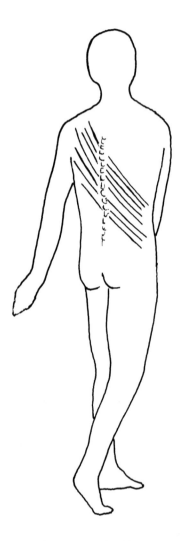

Fig. 10–7. When the trunk is rotated, the back muscles are active on both sides of the spine, with ipsilateral contractions of the spinotransversal muscles and contralateral contractions of the transversospinal muscles. The abdominal muscles are also involved, although their activity is slighter; ipsilateral contractions of the obliquus internus muscle and contralateral contractions of the obliquus externus muscle take place.

The spinotransversal and transversospinal systems of the erector spinae muscles, as well as the abdominal muscles, are active in lateral flexion of the spine. The motion is initiated by ipsilateral contractions of these muscles and modified by contralateral contractions.

Rotation is consistently combined with thoracic lateral flexion. In this combined motion, which is most marked in the upper thoracic spine, the vertebral body rotates in general toward the concavity of the lateral curve of the spine (White, 1969). A combined pattern of rotation and lateral flexion also exists in the lumbar spine (Miles and Sullivan, 1961). In this region the vertebral body rotates toward the convexity of the curve.

Rotation occurs both in the thoracic spine and at the lumbosacral level. Lumbar rotation, except at the lumbosacral level, is quite modest because of the orientation of the facets. During rotation, back and abdominal muscles are active on both sides of the spine, as both ipsilateral and contralateral muscles cooperate (Fig. 10–7). Pelvic motion is essential to increase the range of functional rotation of the trunk.

Functional trunk movements not only are a combination of motion of different parts of the spine, but also depend on a cooperating pelvis. Restriction of motion at any level may increase motion at another level. Thus, a brace worn to restrict thoracic and lumbar motion may result in a transfer of motion to the lumbosacral level (Norton and Brown, 1957; Lumsden and Morris, 1968). Braces and corsets can also affect muscle activity. If a tight brace or corset is worn, the activity of the abdominal muscles is decreased because the brace or corset takes over the function of these muscles if the force exerted on the anterior aspect of the trunk is great enough to support the abdomen (Waters and Morris, 1970).

KINETICS

Loads on the spine are produced primarily by body weight, muscle activity, and externally applied loads. The loads at different levels of the spine can be roughly calculated by using the simplified free body technique. A more accurate calculation of the load on an intervertebral disc can be made from in vitro and in vivo measurements of the pressure within the discs. Since the lumbar region is the main load-bearing area of the spine and the area where pain most commonly occurs, loads on this region are of greatest interest.

Statics of the spine analyzes the loads acting on the spine in equilibrium. These loads differ according to the position of the body. In this section, loads on the lumbar spine during common positions will be examined; loads on the lumbar spine during a common activity in which external loads are applied, that of lifting, will also be discussed.

Dynamics of the spine analyzes the loads acting on the spine during motion. Almost all motion in the body increases the loads on the lumbar spine, from a slight increase during slow walking to a great increase during vigorous physical activity. One area of dynamics of particular interest, that of the loads on the lumbar spine during common strengthening exercises for the back and abdominal muscles, will be discussed.

Standing Posture

The spine can be considered a modified elastic rod because of the behavior of the discs, the longitudinal ligaments, and the elastic ligamentum flavum. The curvatures of the spine in the sagittal plane (kyphosis and lordosis) also contribute to the spring-like capacity of the spine and allow the vertebral column to withstand higher loads than if it were straight. The capacity of the thoracolumbar spine, devoid of muscles, to resist vertical loads has been studied in autopsy spines. The critical load, when buckling occurred, was measured to be about 20 newtons (Lucas and Bresler, 1961). The importance of the extrinsic support of the trunk muscles in stabilizing the spine in the living human being, not only during motion but also in a given position, is obvious.

Postural muscles are always active in the standing position, although their activity is kept at a minimum when the body segments are in good alignment. The center of gravity of the upper part of the body is anterior to the spine. The line of gravity for the trunk most often passes ventral to the center of the fourth lumbar vertebral body (Asmussen and Klausen, 1962). This means that the line of gravity falls ventral to the transverse axis of motion at all spinal levels, subjecting the motion segments to forward-bending moments which must be counterbalanced by ligament forces and back muscle forces (Fig. 10–8).

Even when the body is well balanced, standing is not a completely static position. Any displacement of the line of gravity creates a moment. For the body to remain in equilibrium, this moment must be counteracted by muscle activity, and postural sway therefore occurs intermittently. Not only the erector spinae muscles, but also the abdominal muscles are often intermittently active to maintain the upright position of the trunk. The vertebral portion of the psoas muscle has also been shown to be involved (Basmajian, 1958; Nachemson, 1966).

The level of activity in the different muscle groups varies considerably among individuals. The magnitude of habitual kyphosis and lordosis, for example, influences an individual's posture. Furthermore, the spine cannot be considered in isolation; the pelvis must also be taken into account.

The base of the sacrum is inclined forward and downward, about 30 degrees to the transverse plane during relaxed standing (Fig. 10–9B). This inclination is partly compensated for by the wedge-shaped lumbosacral disc; highest on the ventral side, this disc gives the lowest lumbar vertebra a less inclined position. Tilting of the pelvis about the transverse axis between the hip joints changes the sacral angle. The angle decreases when the pelvis is tilted backward, and the lumbar lordosis flattens (Fig. 10–9A). The

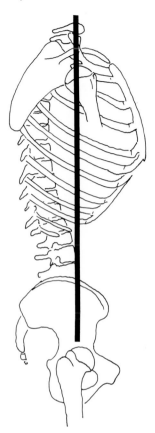

Fig. 10–8.   The line of gravity in relation to the spine is usually ventral to the transverse axis of
motion, and thus the spine is subjected to a constant forward-bending moment.

flattening of the lumbar lordosis affects the thoracic spine, which extends
slightly to adjust the center of gravity. Correspondingly, when the pelvis is
tilted forward the sacral angle increases, causing an increase in the lumbar
lordosis and the thoracic kyphosis (Fig. 10–9C). Changes in the pelvic
inclination influence the activity of the postural muscles by affecting the
statics of the spine. A voluntarily increased pelvic inclination produces
increased back muscle activity, and a decreased inclination diminishes this
activity (Floyd and Silver, 1955).

Effect of Body Position on Loads on the Lumbar Spine

Since the lumbar spine is the main load-bearing part of the vertebral
column, spinal loads have mainly been calculated for this region. The in
vivo disc pressure during relaxed upright standing is a function of the

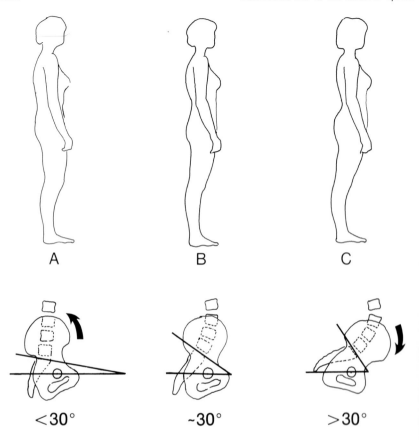

Fig. 10–9. Effect of pelvic tilting on the sacral angle during upright standing.
  A. Backward tilting of the pelvis decreases the sacral angle.
  B. During relaxed standing the sacral angle is about 30 degrees.
  C. Forward tilting of the pelvis increases the sacral angle.
  A decrease in the sacral angle flattens the lumbar spine **(A)**, and an increase in the angle increases the lumbar lordosis **(C)**.

intrinsic pressure in the disc, the body weight above the measured level, and the muscles acting over the motion segment. In a man weighing 70 kg, the load on the third lumbar disc, calculated from the disc pressure, was shown to be 70 kg, that is, almost twice the weight of the body above the measured level, which is about 40 kg (approximately 60% of the total body weight) (Nachemson and Morris, 1964; Nachemson and Elfström, 1970; Nachemson, 1975). Flexion of the trunk increases the load by increasing the forward-bending moment. The forward inclination of the spine makes the disc bulge on the concave side of the spinal curve and contract on the convex side (Fig. 10–10). Thus, in flexion the disc protrudes anteriorly and is depressed posteriorly. Both compressive and tensile stresses increase. If

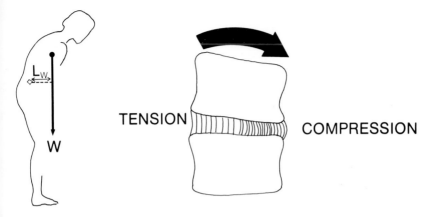

Fig. 10–10. The forward-bending position produces a bending moment on the lumbar spine. The moment is a product of the force produced by the weight of the upper body (W) and the lever arm of the force (L_w). The forward inclination of the upper body subjects the disc to increased tensile and compressive stresses. The disc bulges on the compressive side and is depressed on the tensile side.

rotatory motion and thus torsional loads are added, the stresses on the disc will increase further (Andersson et al., 1977).

During relaxed unsupported sitting the loads on the lumbar spine are greater than during relaxed upright standing (Nachemson and Elfström, 1970; Andersson et al., 1974). In relaxed unsupported sitting the pelvis is tilted backward and the lumbar lordosis is straightened out. The line of gravity for the upper body, already ventral to the lumbar spine, shifts further ventrally, creating a longer lever arm for the force exerted by the weight of the trunk (Fig. 10–11A and B). This longer lever arm produces an increased torque in the lumbar spine; this torque increases further if the trunk is bent forward. Moreover, the activity of the psoas muscle has been found to contribute to the loads on the lumbar region (Nachemson, 1968). During erect sitting, forward tilting of the pelvis and an increase in the lumbar lordosis reduce the loads on the lumbar spine, but these loads still exceed those produced during relaxed upright standing (Fig. 10–11C). The relative loads on the spine in various body positions are presented in Figure 10–12.

During supported sitting the loads on the lumbar spine are less than during unsupported sitting because part of the weight of the upper body is supported by the backrest. A backward inclination of the backrest, as well as the use of a lumbar support, further decreases the loads. However, a support in the thoracic region which pushes the thoracic spine and the trunk forward makes the lumbar spine move toward kyphosis to remain in contact with the backrest, increasing the loads on the lumbar spine (Andersson et al., 1974) (Fig. 10–13).

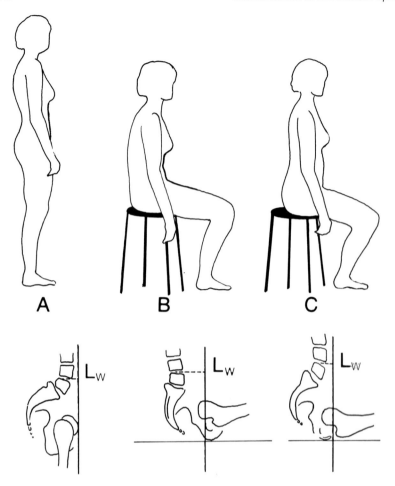

Fig. 10–11. Compared with relaxed upright standing **(A)**, the line of gravity for the upper body, already ventral to the lumbar spine, shifts further ventrally during relaxed unsupported sitting as the pelvis is tilted backward and the lumbar lordosis flattens **(B)**. This shift creates a longer lever arm (L_W) for the force exerted by the weight of the upper body. During erect sitting the backward pelvic tilt is reduced and the lever arm shortens **(C)**, but it is still slightly longer than during relaxed upright standing.

Fig. 10–13. Influence of backrest inclination and back support on loads on the lumbar spine, in terms of pressure in the third lumbar disc, during supported sitting.
A. Backrest inclination is 90 degrees, and disc pressure is at a maximum.
B. Addition of a lumbar support decreases the disc pressure.
C. Backward inclination of the backrest to 110 degrees, but with no lumbar support, produces less disc pressure.
D. Addition of a lumbar support with this degree of backrest inclination further decreases the pressure.
E. Shifting the support to the thoracic region pushes the upper body forward, moving the lumbar spine toward kyphosis and increasing the disc pressure. (Adapted from Andersson et al., 1974.)

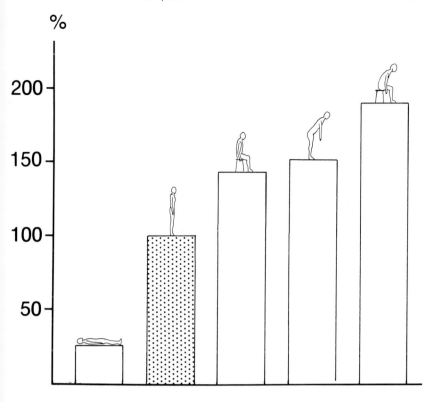

Fig. 10–12. The relative loads on the third lumbar disc for various body positions in living subjects are compared with the load during upright standing, depicted as 100%. (Adapted from Nachemson, 1975.)

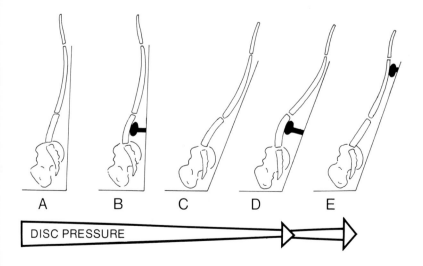

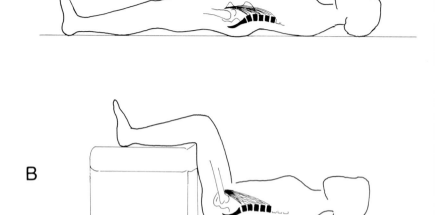

Fig. 10–14.   **A.** In the supine position with straight legs, some loads on the lumbar spine are
produced by the pull of the vertebral portion of the psoas muscle.
   B. With hips and knees bent the psoas muscle relaxes, thus decreasing the load on the
lumbar spine.

Minimum loads on the spine occur in the supine position, where the loads
produced by body weight are eliminated (Fig. 10–12). When the knees are
extended in this position, the pull of the vertebral portion of the psoas
muscle produces some load on the lumbar spine. When the hips and knees
are bent and supported, the lumbar lordosis straightens out as the psoas
muscle is relaxed, and the loads decrease (Fig. 10–14). A further decrease in
the loads is achieved by applying traction. When traction is applied to a
patient in the supine position with hips and knees bent and supported to
flatten the spine, the forces produced by the traction are more evenly
distributed throughout the spine than if the lumbar lordosis is maintained by
keeping the legs straight (Fig. 10–15).

Effect of Lifting on Loads on the Lumbar Spine

When a vertebral body is subjected to a compressive load of a certain
magnitude, a fracture will occur. Studies of lumbar vertebral specimens from
adult humans showed that the compressive load to failure ranged from about
5,000 to 8,000 newtons (Eie, 1966). These values correspond on the whole
to those of other authors, although values of about 10,000 newtons, as well
as below 5,000 newtons, have been documented. Age and degree of disc
degeneration influence this range. The finding that the fracture point was
reached before damage to the intervertebral disc occurred shows that the
bone is less capable of resisting compression than the disc if not predamaged

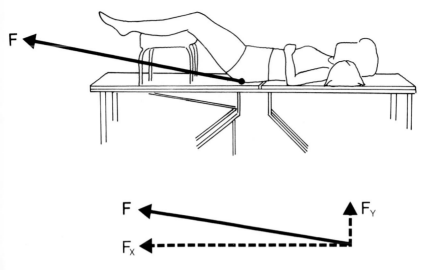

Fig. 10–15. The semiflexed position, which flattens the lumbar spine, favors a more even distribution of the force produced by traction (F). The pull must be directed diagonally to keep the spine flat. This diagonal arrangement means that not all of the pull exerts a horizontal force. Resolving the traction force (F) into horizontal (F_X) and vertical (F_Y) components demonstrates that part of the pull produces lifting. In this example friction forces are not taken into account.

(Eie, 1966). Before the vertebra or the end-plate fractures, a yield point is reached. If the load is removed at this point, the vertebral body recovers but may be more susceptible to damage if reloaded. More recent studies have shown the presence of microfractures in specimens from "normal" lumbar vertebrae (Hansson, 1977). This microdamage may be interpreted to be fatigue fractures resulting from stresses and strains on the spine in vivo. Correspondingly, radiating ruptures in the posterior part of the annulus fibrosus may be a result of excessive tension from high bending and torsional loads which are dangerous to the disc.

Lifting and carrying an object are common situations wherein an external load is applied to the vertebral column. Several factors influence the loads on the spine during these activities:

—The position of the object in relation to the center of motion in the spine
—The degree of flexion or rotation of the spine
—The characteristics of the object: size, shape, weight, and density

Holding the load close to the body instead of away from it decreases the bending moment on the lumbar spine because the distance between the center of gravity for the object and the center of motion in the spine is minimized. The shorter the lever arm is for the force produced by the weight of a given object, the lower the magnitude of the bending moment, and thus the lower the loads on the lumbar spine (Andersson, Örtengren, and Nachemson, 1976).

The size and the shape of the object carried, as well as its weight and density, influence the loads on the spine. If objects of equal weight, the same shape, and a uniform density but of different sizes are held, the lever arm for the force produced by the weight of the object will be longer for the larger

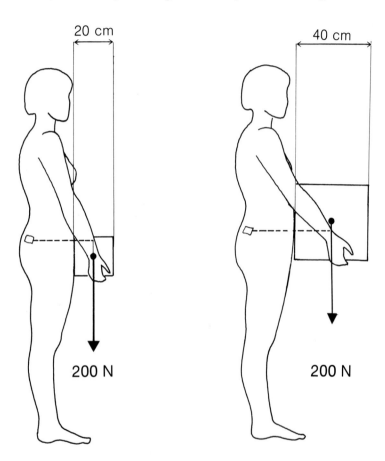

A. FORWARD-BENDING MOMENT = 60 Nm

B. FORWARD-BENDING MOMENT = 80 Nm

Fig. 10–16. The size of the object carried influences the loads on the lumbar spine. In two situations the distance from the center of motion in the disc to the front of the abdomen is 20 cm. In both cases the object has a uniform density and weighs 20 kg. In Case A the length of the cubic object is 20 cm; in Case B the length is 40 cm. Thus, in Case A the forward-bending moment acting on the lowest lumbar disc is 60 newton meters, as the force of 200 newtons produced by the weight of the object acts with a lever arm (L_p) of 30 cm (200 N × 0.3 m). In Case B the forward-bending moment is 80 newton meters, as the lever arm (L_p) is 40 cm (200 N × 0.4 m).

object, and thus the bending moment in the lumbar spine will be greater (Fig. 10–16).

When an object is held with the body in a forward-bending position, not only the force produced by the weight of the object, but also that produced by the weight of the upper body will create a bending moment on the disc, resulting in an increased load on the spine. This bending moment will be greater than that produced during upright standing (Fig. 10–17).

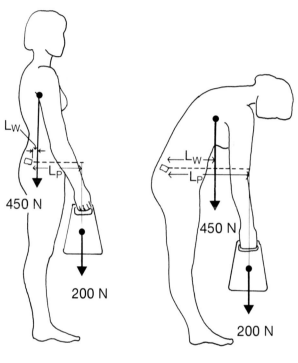

A. TOTAL FORWARD-
BENDING MOMENT
= 69 Nm

B. TOTAL FORWARD-
BENDING MOMENT
= 192.5 Nm

Fig. 10–17. The position of the upper body during lifting influences the loads on the lumbar spine. In two situations an identical object weighing 20 kg is lifted. In Case A (upright standing), the lever arm of the force produced by the weight of the object (L_p) is 30 cm, creating a forward-bending moment of 60 newton meters (200 N × 0.3 m). The forward-bending moment of the upper body is 9 newton meters. The length of the lever arm (L_w) is estimated to be 2 cm, and the force produced by the weight of the upper body is 450 newtons. Thus, the total forward-bending moment in Case A is equal to 69 newton meters (60 Nm + 9 Nm). In Case B (upper body flexed forward), the lever arm of the force produced by the weight of the object (L_p) is increased to 40 cm, creating a forward-bending moment of 80 newton meters (200 N × 0.4 m). Furthermore, the force of 450 newtons produced by the weight of the upper body becomes important, as it acts with a lever arm (L_w) of 25 cm, creating a forward-bending moment of 112.5 newton meters (450 N × 0.25 m). Thus, the total forward-bending moment in Case B is 192.5 newton meters (112.5 Nm + 80 Nm).

During lifting, it is generally recommended that the work be carried out with knees bent to reduce the load on the spine. This recommendation is only valid if the technique is correctly employed. Lifting with bent knees makes it possible to hold the object (if not too big) closer to the trunk and consequently closer to the center of motion in the spine (Fig. 10–18A and B). However, the loads are not reduced if the object is held out in front of the knees, that is, farther away from the center of motion; despite the bent knees, lifting the load in this way increases rather than decreases the bending moment (Fig. 10–18C).

The loads on the spine at any point during lifting of an object can be roughly calculated using the free body technique. A theoretical example is the calculation of loads on a lumbar disc at one point in time during lifting of an object weighing 20 kg by an individual weighing 70 kg. Flexion of the

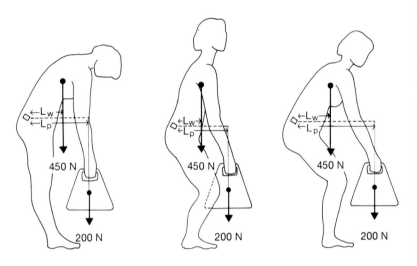

| A. TOTAL FORWARD-BENDING MOMENT = 192.5 Nm | B. TOTAL FORWARD-BENDING MOMENT = 151 Nm | C. TOTAL FORWARD-BENDING MOMENT = 212.5 Nm |

Fig. 10–18. The technique employed during lifting influences the loads on the lumbar spine. In three situations an identical object weighing 20 kg is lifted. Case A (upper body flexed forward) is identical to Case B in Figure 10–17; the total forward-bending moment is 192.5 newton meters. In Case B, lifting with knees bent and back straight places the object closer to the trunk, decreasing the forward-bending moments. The lever arms of the forces produced by the weight of the object (L_p) and upper body (L_w) are shortened to 35 and 18 cm, respectively, at this point in the lifting process. The result is a total forward-bending moment of 151 newton meters ([200 N × 0.35 m] + [450 N × 0.18 m]). Case C shows that bent knees per se do not decrease the forward-bending moments. If the object lifted is held out in front of the knees, the lever arm of the force produced by the weight of the object (L_p) increases to 50 cm, and the lever arm of the force produced by the weight of the upper body (L_w) increases to 25 cm. Thus, the total forward-bending moment created is 212.5 newton meters ([200 N × 0.5 m] + [450 N × 0.25 m]).

spine is about 35 degrees. In this example (Fig. 10–19A), the three main forces acting on the lumbar spine at the lumbosacral level are:

1. The force produced by the weight of the upper body (W), calculated as 450 newtons (approximately 65% of the force exerted by the total body weight)

2. The force produced by the weight of the object (P), 200 newtons

3. The force produced by contraction of the erector spinae muscles (M), which has a known direction and point of application but an unknown magnitude

Because these three forces are located away from the instantaneous center of motion in the spine, they create moments in the lumbar spine. Two forward-bending moments (moments WL_W and PL_P) are produced by force W and force P and their distances from the instant center. A counterbalancing moment (moment ML_M) is produced by force M and its distance from the instant center. The lever arm for force W (the perpendicular distance of the force from the instant center) is 0.25 meters. The lever arm for force P is 0.4 meters, and for force M, 0.05 meters. The magnitude of force M can be found by using an equilibrium equation. For the body to be in equilibrium, the sum of the moments acting on the lumbar spine must be zero. (In this example, moments acting clockwise are considered to be positive, while counterclockwise moments are considered to be negative.)

Thus,
$$\Sigma \text{ moments} = 0$$
$$(W \times L_W) + (P \times L_P) - (M \times L_M) = 0$$
$$(450 \text{ N} \times 0.25 \text{ m}) + (200 \text{ N} \times 0.4 \text{ m}) - (M \times 0.05 \text{ m}) = 0$$
$$M \times 0.05 \text{ m} = 112.5 \text{ Nm} + 80 \text{ Nm}.$$

Solving this equation for M yields 3,850 newtons, which is the value of force M.

The total compressive force exerted on the disc (force C) can now be calculated by using trigonometric methods (Fig. 10–19B). If the disc is inclined 35 degrees to the transverse plane, force C in this example must be the sum of the compressive forces acting over the disc. These forces are:

1. The compressive force produced by the weight of the upper body (W), which acts on the disc inclined 35 degrees (W × cos 35)

2. The force produced by the weight of the object (P), which acts on the disc inclined 35 degrees (P × cos 35)

3. The force produced by the erector spinae muscles (M), which acts approximately at right angles to the disc inclination

The total compressive force acting on the disc (C) has a known sense, point of application, and line of application, but an unknown magnitude. The magnitude of force C can be found by using an equilibrium equation. For the body to be in equilibrium, the sum of the forces must equal zero.

Thus,
$$\Sigma \text{ forces} = 0$$
$$(W \times \cos 35) + (P \times \cos 35) + M - C = 0$$
$$(450 \text{ N} \times \cos 35) + (200 \text{ N} \times \cos 35) + 3850 \text{ N} - C = 0$$
$$C = 368.5 \text{ N} + 163.8 \text{ N} + 3850 \text{ N}.$$

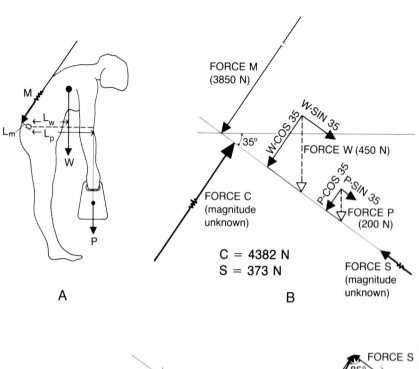

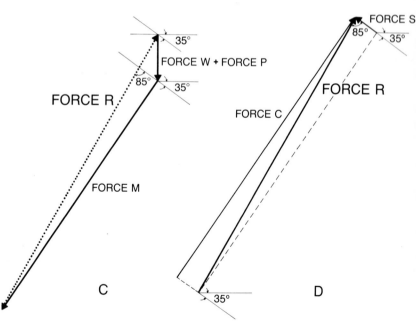

Solving the equation for C yields 4,382 newtons. For solutions of the shear force and the total reaction force acting on the disc, see Figures 10–19B, C, and D.

Calculations made in this way for one point in time during lifting are valuable for understanding the importance of the lever arms of the forces produced by the weight of the upper body and the object for the loads that are imposed on the spine. However, the values cannot represent the actual load on the lumbar spine in vivo, particularly when heavier objects are involved. Using the same calculations to compute the loads produced when an object of 80 kg is lifted (representing a force of 800 newtons) should give an approximate load on the disc of 10,000 newtons, which exceeds the fracture point of the vertebra.

Fig. 10–19. **A.** Three main forces act on the lumbar spine at the lumbosacral level:
1. The force produced by the weight of the upper body (W), 450 newtons
2. The force produced by the weight of the object (P), 200 newtons
3. The force produced by the erector spinae muscles (M), magnitude unknown

Two forward-bending moments, moments WL_W and PL_P, are produced by forces W and P and their distances from the instant center. The lever arm (L_P) for force P is 40 cm, and the lever arm (L_W) for force W is 25 cm. The counterbalancing moment, moment ML_M, is produced by force M and its distance from the instant center. The lever arm (L_M) is 5 cm. The magnitude of force M is found by solving the equilibrium equation which states that the sum of all the moments must equal zero:
$$(450 \text{ N} \times 0.25 \text{ m}) + (200 \text{ N} \times 0.4 \text{ m}) - (M \times 0.05 \text{ m}) = 0$$
$$M \times 0.05 \text{ m} = 112.5 \text{ Nm} + 80 \text{ Nm}$$
$$M = 3850 \text{ N}.$$
B. Forces W and P are resolved into a compression component, $W \times \cos 35$ and $P \times \cos 35$ respectively, and a shear component, $W \times \sin 35$ and $P \times \sin 35$ respectively. The total compressive reaction force acting on the disc (C) can now be solved by using trigonometric functions. The magnitude of force C is found by solving the equation which states that the sum of all the forces must equal zero:
$$(450 \text{ N} \times \cos 35) + (200 \text{ N} \times \cos 35) + 3850 \text{ N} - C = 0$$
$$C = 4382 \text{ N}.$$
The shear component for the reaction force on the disc (S) is solved in the same way:
$$(450 \text{ N} \times \sin 35) + (200 \text{ N} \times \sin 35) - S = 0$$
$$S = 373 \text{ N}.$$
Since force C and force S form a right angle, the Pythagorean theorem can be used to find the total reaction force on the disc (R):
$$R = \sqrt{J^2 + S^2}$$
$$R = 4398 \text{ N}.$$
The direction of force R ($\sqrt{}$) is determined by using one of the trigonometric functions:
$$\sin \sqrt{} = \frac{C}{R}$$
$$\sqrt{} = 85°$$
Thus, the direction forms an 85-degree angle with the inclined disc.
C. The problem can be graphically solved by constructing a vector diagram based on the known values. Starting with W + P which are vertical, M is added at a right angle to the inclined disc, and R closes the triangle. The direction of R in relation to the disc is measured.
D. When the magnitude and direction of the reaction force R are known, R can be resolved into its compressive force component C, and its shearing component S, by drawing a parallelogram with R as the diagonal. Force S, which mainly represents the shear-resisting effect of the disc and the posterior elements of the motion segment, is proportionately rather small in this example but will increase with increasing inclination of the disc.

Since athletes while lifting can easily reach such calculated loads without sustaining fractures, other factors must be involved in reducing the loads on the spine. The effect of the intra-abdominal support in reducing the loads on the lumbosacral spine has been calculated from intra-abdominal pressure measurements (Bartelink, 1957; Morris et al., 1961; Eie and Wehn, 1962). Using calculations of the abdominal cross-sectional area and the perpendicular distance of the muscular wall from the fifth spinous process, Eie and Wehn (1962) converted the abdominal pressure to a moment to be substituted into an equation for the different torques acting on the disc. The intra-abdominal pressure in athletes was shown to reduce the loads on the spine produced by erector spinae contraction by up to 40%.

During moderate lifting, reduction of the loads on the spine due to intra-abdominal support has not been demonstrated to reach values of such a magnitude. On the contrary, the effect has been felt to be negligible when static positions are involved or when the spine is in moderate flexion (Asmussen and Poulsen, 1968; Andersson, Örtengren, and Nachemson, 1976). However, in most studies there is agreement that the intra-abdominal pressure increases with increasing forward bending of the trunk and with increased weight of the load lifted. Higher values are also found during dynamic lifting, when acceleration forces are involved, than at one static instant during the lift. Particularly in the initial phase of lifting, just at the point when the weight of the load is about to be overcome, a large increase is found in the intra-abdominal pressure, which probably plays a major part in stabilizing the spine and moderating the intervertebral compressive forces. The muscular activity needed to build up the pressure is mainly provided by the transverse and oblique abdominal muscles.

During lifting, the erector spinae muscles are more or less active, depending on the degree of flexion of the trunk. Their activity increases as the bending moment increases within a certain range of motion, that is, from the upright standing position to about 60 degrees of flexion. In any lifting situation, however, the bending moment must be counterbalanced by opposite forces.

The muscles or the ligaments may be involved to a varying degree when different lifting techniques are used. A static pull of a given magnitude from a small distance from the floor requires less back muscle activity when the back-lifting method is used (knees straight and trunk flexed), with the trunk flexed more than 60 degrees (Fig. 10–20A), than when the leg-lifting method is used (knees bent and back straight) (Fig. 10–20B) (Andersson, Herberts, and Örtengren, 1976). This does not mean, however, that the back-lifting method creates less load on the lumbar spine at this point in the lifting process. Although less muscle activity is needed to counterbalance the upper body when the trunk is flexed more than 60 degrees, an increase in tension is needed in the ligaments at this point to maintain the body position (Farfan, 1975). The stretched erector spinae muscles and the posterior hip muscles also help to maintain the position of the body.

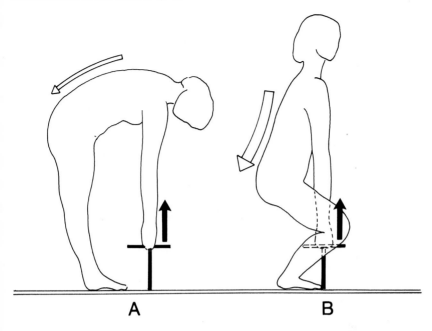

Fig. 10–20. In two static pulls of the same magnitude from a small distance from the floor, the activity of the erector spinae muscles (designated by the hollow arrows) varies with the lifting technique employed. In Case A (knees straight and upper body flexed more than 60 degrees), the back muscle activity is less than in Case B (knees bent and back straight). However, this does not mean that less load on the spine is produced in Case A, because in this situation the forward-bending moment is also counterbalanced by ligament forces. In Case B, the ligaments are slack and the counterbalancing force is mainly provided by muscle contraction; thus, the ligaments are protected from excessive strain. On the other hand, the deep bending of the knees subjects the knee joints to heavy loads and increases the demand on the leg muscles, which are in an unfavorable position for exerting force to raise the body.

At the beginning of the lifting process, the pelvis tilts backward and lumbar extension is delayed; thus, back muscle activity is low during this phase of lifting (Davis et al., 1965). When the spine starts to extend, the muscles increase their activity and the forward-bending moment is counteracted by the musculature. Since the forces produced by the ligaments have shorter lever arms than those produced by the muscles, the load on the ligaments may become extremely high if stabilization of the spine is not provided by muscle contractions. Thus, to protect the ligaments and the spine, back muscles should be active when lifting is begun, but no motion should occur within the spine until the initial phase is over, that is, until the inertia of the load is overcome. Spinal motion at the very start of lifting increases the stresses on the motion segment.

The resistance of the lumbar spine to bending forces is lower than to compressive forces. In vitro studies have shown that greater resistance to failure is provided when the loads are applied centrally than when applied eccentrically or at an inclination (Lin et al., 1978). This finding suggests that a vertical position of the spine during lifting to prevent wedging of the discs is preferred. In vivo measurements have also demonstrated that an increased disc pressure is produced in the lumbar spine when the trunk is loaded in lateral flexion or in flexion combined with rotation (Andersson et al., 1977).

Effect of Exercises on Loads on the Lumbar Spine

Almost all motion in the body increases the loads on the lumbar spine, from a modest increase during, for example, slow walking or easy twisting to a more marked increase during some exercises (Nachemson and Elfström, 1970). An area of dynamics which is of great interest is the loads on the spine during strengthening exercises for the erector spinae and abdominal muscles. Such exercises should be effective for strengthening the muscles concerned but should also be performed in such a way that the loads produced on the spine are adjusted to the condition of the individual's back.

Arching the back in the prone position, as illustrated in Figure 10–21A, greatly activates the erector spinae muscles (Pauly, 1966). But because loading of the spine in extreme positions produces greater stresses on the spinal structures than does more centrally applied loading, this hyperextended position should be avoided. Therefore, when strengthening exercises for the erector spinae musculature are performed, an initial position that keeps the vertebrae more parallel is preferable (Fig. 10–21B).

Bilateral straight-leg raising is commonly used as an abdominal muscle strengthening exercise. This exercise, however, does little to activate the abdominal muscles. Instead, the vertebral portion of the psoas muscle is most active and tends to pull the lumbar spine into lordosis. Performing a sit-up from the supine position, with hips and knees bent to limit the psoas activity, effectively activates the abdominal muscles but also creates a great increase in the lumbar disc pressure (Nachemson and Elfström, 1970). The load on the lumbar spine is lessened if the range of motion of the exercise is limited by performing a trunk curl, wherein the head and shoulders are only raised to the point where the shoulder blades clear the table, excluding lumbar spinal motion (Fig. 10–22). This modification of the exercise has been shown to be quite effective in terms of motor unit recruitment in the muscles (Partridge and Walters, 1959; Flint, 1965); all portions of the external oblique and rectus abdominis muscles are activated. A reverse curl, wherein the knees are brought toward the chest and the buttocks raised from the table, activates the internal and external oblique muscles and the rectus abdominis muscle (Partridge and Walters, 1959). If the reverse curl is performed isometrically, the disc pressure will be lower than that produced by a sit-up, but the exercise will be just as effective for strengthening the abdominal muscles (Fig. 10–23).

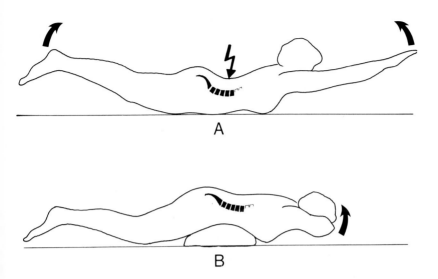

Fig. 10–21. **A.** Arching the back in the prone position greatly activates the erector spinae muscles but also produces high stresses on the lumbar discs, which are loaded in an extreme position.
    B. Decreasing the arch of the back by placing a pillow under the abdomen allows the discs to better resist stresses, since the vertebrae are kept more parallel to each other. Isometric exercise is preferable in this position.

Fig. 10–22. Performing a curl to the point where only the shoulder blades clear the table excludes lumbar motion, and the load on the lumbar spine is less than when a full sit-up is performed. A greater moment is produced if the arms are raised above the head or the hands are clasped behind the neck, as the center of gravity of the upper body then shifts farther away from the center of motion.

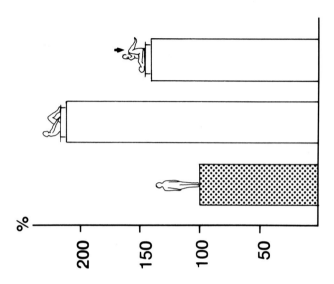

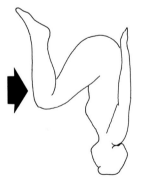

Fig. 10–23. A reverse curl, isometrically performed, provides efficient training of the abdominal muscles and produces moderate stresses on the lumbar discs. The relative loads on the third lumbar disc for a full sit-up and an isometric curl are compared with the load during upright standing, depicted as 100%. (Adapted from Nachemson, 1975.)

SUMMARY

1. A vertebra-disc-vertebra unit constitutes a motion segment—the functional unit of the spine. If subjected to compressive loads of a certain magnitude, the end-plate or the vertebral body will fracture before the disc is damaged.
2. When compressive loads are imposed on the disc, the compressive stress is highest in the nucleus pulposus, whereas tensile stress predominates in the surrounding structure—the annulus fibrosus.
3. During bending the disc bulges on the concave side of the spinal curve and contracts on the convex side. Bending and torsional loads create higher stresses on the disc than do axial compressive loads. The ability of the disc to store energy and to distribute loads is reduced with degeneration of the disc.
4. The intervertebral joints may sustain a certain amount of compressive load depending upon the position of the body. In addition, they guide the motion of the motion segment. The orientation of the facets determines the type of motion.
5. The instantaneous center of motion usually lies within the disc. Motion between two vertebrae is small and does not occur independently in vivo. Thus, functional motion of the spine is always a combined action of several segments. The range of motion varies among individuals and decreases with age.
6. Trunk motion involves pelvic motion. The first 50 to 60 degrees of flexion occurs in the lumbar spine, and further flexion is mainly produced by forward tilting of the pelvis.
7. The trunk muscles involved in motion also play an important role in providing extrinsic stability to the spine; intrinsic stability is provided by the ligaments and discs. Maintaining an upright standing position requires muscular activity.
8. The loads on the lumbar disc are relatively high in vivo. During upright standing the compressive load is about twice the superimposed body weight.
9. Body position affects the loads on the lumbar spine. A forward-bending or twisted position causes higher stresses on the lumbar spine than does upright standing. Unsupported sitting also produces higher loads than does upright standing.
10. Externally applied loads, produced, for example, by lifting or carrying objects, may subject the lumbar spine to very high loads, although the intra-abdominal pressure helps to support the spine. To reduce the load on the spine during lifting, the distance between the trunk and the object lifted should be as short as possible.
11. The motion segment has a greater resistance to failure if loaded centrally than if loaded eccentrically.
12. The ligamentous structures, which have a limited resiliency, are protected from excessive strain by muscular contraction.

REFERENCES

Andersson, G. B. J., Herberts, P., and Örtengren, R.: Myoelectric back muscle activity in standardized lifting postures. In *International Series on Biomechanics.* Vol. 1A. Edited by P. V. Komi. Baltimore, University Park Press, 1976, pp. 520–529.

Andersson, G. B. J., Örtengren, R., and Nachemson, A.: Quantitative studies of back loads in lifting. Spine, *1*:178, 1976.

————: Intradiskal pressure, intra-abdominal pressure and myoelectric back muscle activity related to posture and loading. Clin. Orthop., *129*:156, 1977.

Andersson, G. B. J., Örtengren, R., Nachemson, A., and Elfström, G.: Lumbar disc pressure and myoelectric back muscle activity during sitting. I. Studies on an experimental chair. Scand. J. Rehabil. Med., 6:104, 1974.

Asmussen, E., and Klausen, K.: Form and function of the erect human spine. Clin. Orthop., *25*:55, 1962.

Asmussen, E., and Poulsen, E.: On the role of the intra-abdominal pressure in relieving the back muscles while holding weights in a forward inclined position. Comm. Dan. Nat. Assoc. Infant. Paral., *28*:3, 1968.

Bartelink, D. L.: The role of abdominal pressure in relieving the pressure on the lumbar intervertebral discs. J. Bone Joint Surg., *39B*:718, 1957.

Basmajian, J. V.: Electromyography of iliopsoas. Anat. Rec., *132*:127, 1958.

Carlsöö, S.: The static muscle load in different work positions: An electromyographic study. Ergonomics, *4*:193, 1961.

Cossette, J. W., Farfan, H. F., Robertson, G. H., and Wells, R. V.: The instantaneous center of rotation of the third lumbar intervertebral joint. J. Biomech., *4*:149, 1971.

Davis, P. R., Troup, J. D. G., and Burnard, J. H.: Movements of the thoracic and lumbar spine when lifting: A chronocyclophotographic study. J. Anat. Lond., 99:13, 1965.

Eie, N.: Load capacity of the low back. J. Oslo City Hosp., *16*:73, 1966.

Eie, N., and Wehn, P.: Measurements of the intra-abdominal pressure in relation to weight bearing of the lumbosacral spine. J. Oslo City Hosp., *12*:205, 1962.

Farfan, H. F.: Muscular mechanism of the lumbar spine and the position of power and efficiency. Orthop. Clin. North Am., 6:135, 1975.

Flint, M. M.: Abdominal muscle involvement during the performance of various forms of sit-up exercise. An electromyographic study. Am. J. Phys. Med., *44*:224, 1965.

Floyd, W. F., and Silver, P. H. S.: The function of the erectores spinae muscles in certain movements and postures in man. J. Physiol., *129*:184, 1955.

Friedebold, G.: Die Aktivität normaler Rückenstreckmuskulatur im Elektromyogramm unter verschiedenen Haltungsbedingungen; eine Studie zur Skelettmuskelmechanik. Z. Orthop., *90*:1, 1958.

Galante, J. O.: Tensile properties of the human lumbar annulus fibrosus. Acta Orthop. Scand., Suppl. *100*:1–91, 1967.

Hansson, T.: The bone mineral content and biomechanical properties of lumbar vertebrae. An *in vitro* study based on dual photon absorptiometry. Thesis, University of Göteborg, Sweden, 1977.

Joseph, J.: *Man's Posture: Electromyographic Studies.* Springfield, Charles C Thomas, 1960.

King, A. I., Prasad, P., and Ewing, C. L.: Mechanism of spinal injury due to caudocephalad acceleration. Orthop. Clin. North Am., 6:19, 1975.

Kulak, R. F., Schultz, A. B., Belytschko, T., and Galante, J.: Biomechanical characteristics of vertebral motion segments and intervertebral discs. Orthop. Clin. North Am., 6:121, 1975.

Lin, H. S., Liu, Y. K., and Adams, K. H.: Mechanical response of the lumbar intervertebral joint under physiological (complex) loading. J. Bone Joint Surg., *60A*:41, 1978.

Lucas, D. B., and Bresler, B.: Stability of the ligamentous spine. Biomechanics Laboratory, University of California, San Francisco and Berkeley. Technical Report 40. San Francisco, The Laboratory, 1961.

Lumsden, R. M., and Morris, J. M.: An *in vivo* study of axial rotation and immobilization at the lumbosacral joint. J. Bone Joint Surg., *50A*:1591, 1968.

Miles, M., and Sullivan, W. E.: Lateral bending at the lumbar and lumbosacral joints. Anat. Rec., *139*:387, 1961.

Moll, J. M. H., and Wright, V.: Normal range of spinal mobility. An objective clinical study. Ann. Rheum. Dis., *30*:381, 1971.

Morris, J. M., Lucas, D. B., and Bresler, B.: Role of the trunk in stability of the spine. J. Bone Joint Surg., *43A*:327, 1961.

Nachemson, A.: Lumbar intradiscal pressure. Acta Orthop. Scand., Suppl. *43*:1–140, 1960.
———: The influence of spinal movements on the lumbar intradiscal pressure and on the tensile stresses in the annulus fibrosus. Acta Orthop. Scand., *33*:183, 1963.
———: Electromyographic studies on the vertebral portion of the psoas muscle. With special reference to its stabilizing function of the lumbar spine. Acta Orthop. Scand., *37*:177, 1966.
———: The possible importance of the psoas muscle for stabilization of the lumbar spine. Acta Orthop. Scand., *39*:47, 1968.
———: Towards a better understanding of back pain; a review of the mechanics of the lumbar disc. Rheumatol. Rehabil., *14*:129, 1975.
Nachemson, A., and Elfström, G.: *Intravital Dynamic Pressure Measurements in Lumbar Discs: A Study of Common Movements, Maneuvers and Exercises.* Stockholm, Almqvist & Wiksell, 1970.
Nachemson, A. L., and Evans, J. H.: Some mechanical properties of the third human lumbar interlaminar ligament (ligamentum flavum). J. Biomech., *1*:211, 1968.
Nachemson, A., and Morris, J. M.: *In vivo* measurements of intradiscal pressure. Discometry, a method for the determination of pressure in the lower lumbar discs. J. Bone Joint Surg., *46A*:1077, 1964.
Norton, P. L., and Brown, T.: The immobilizing efficiency of back braces. Their effect on the posture and motion of the lumbosacral spine. J. Bone Joint Surg., *39A*:111, 1957.
Partridge, M. J., and Walters, C. E.: Participation of the abdominal muscles in various movements of the trunk in man. An electromyographic study. Phys. Ther. Rev., *39*:791, 1959.
Pauly, J. E.: An electromyographic analysis of certain movements and exercises. I. Some deep muscles of the back. Anat. Rec., *155*:223, 1966.
Reichmann, S.: Motion of the lumbar articular processes in flexion-extension and lateral flexion of the spine. Acta Morphol. Neerl. Scand., *8*:261, 1970/71.
Reichmann, S., Berglund, E., and Lundgren, K.: Das Bewegungszentrum in der Lendenwirbelsäule bei Flexion und Extension. Z. Anat. Entwicklungsgesch., *138*:283, 1972.
Rolander, S. D.: Motion of the lumbar spine with special reference to the stabilizing effect of posterior fusion. An experimental study on autopsy specimens. Acta Orthop. Scand., Suppl. *90*:1–144, 1966.
Waters, R. L., and Morris, J. M.: Effect of spinal supports on the electrical activity of muscles of the trunk. J. Bone Joint Surg., *52A*:51, 1970.
White, A. A.: Analysis of the mechanics of the thoracic spine in man. An experimental study of autopsy specimens. Acta Orthop. Scand., Suppl. *127*:1–105, 1969.
White, A. A., and Panjabi, M. M.: *Clinical Biomechanics of the Spine.* Philadelphia, J. B. Lippincott Co., 1978.

SUGGESTED READING

Andersson, G. B. J.: On myoelectric back muscle activity and lumbar disc pressure in sitting postures. Scand. J. Rehabil. Med., Suppl. 3, 1974.
Andersson, G. B. J., Herberts, P., and Örtengren, R.: Myoelectric back muscle activity in standardized lifting postures. In *International Series on Biomechanics.* Vol. 1A. Edited by P. V. Komi. Baltimore, University Park Press, 1976, pp. 520–529.
Andersson, G. B. J., Örtengren, R., and Nachemson, A.: Quantitative studies of back loads in lifting. Spine, *1*:178–185, 1976.
———: Intradiskal pressure, intra-abdominal pressure and myoelectric back muscle activity related to posture and loading. Clin. Orthop., *129*:156–164, 1977.
Andersson, G. B. J., Örtengren, R., Nachemson, A., and Elfström, G.: Lumbar disc pressure and myoelectric back muscle activity during sitting. I. Studies on an experimental chair. Scand. J. Rehabil. Med., *6*:104–114, 1974.
Asmussen, E., and Klausen, K.: Form and function of the erect human spine. Clin. Orthop., *25*:55–63, 1962.
Asmussen, E., and Poulsen, E.: On the role of the intra-abdominal pressure in relieving the back muscles while holding weights in a forward inclined position. Comm. Dan. Nat. Assoc. Infant. Paral., *28*:3–11, 1968.

Asmussen, E., Poulsen, E., and Rasmussen, B.: Quantitative evaluation of the activity of the back muscles in lifting. Comm. Dan. Nat. Assoc. Infant. Paral., *21*:3–14, 1965.
Bartelink, D. L.: The role of abdominal pressure in relieving the pressure on the lumbar intervertebral discs. J. Bone Joint Surg., *39B*:718–725, 1957.
Basmajian, J. V.: Electromyography of iliopsoas. Anat. Rec., *132*:127–132, 1958.
Brown, T., Hansen, R. J., and Yorra, A. J.: Some mechanical tests on the lumbosacral spine with particular reference to the intervertebral discs. A preliminary report. J. Bone Joint Surg., *39A*:1135–1164, 1957.
Carlsöö, S.: The static muscle load in different work positions: An electromyographic study. Ergonomics, *4*:193–211, 1961.
————: *Att lyfta i jobbet.* Uddevalla, Sweden, Pa-radet, Bohusläningens AB, 1975.
Chaffin, D. B., and Baker, W. H.: A biomechanical model for analysis of symmetric sagittal plane lifting. Am. Inst. Indus. Engng. Trans., *2*:16–27, 1970.
Cossette, J. W., Farfan, H. F., Robertson, G. H., and Wells, R. V.: The instantaneous center of rotation of the third lumbar intervertebral joint. J. Biomech., *4*:149–153, 1971.
Davis, P. R., and Troup, J. D. G.: Pressures in the trunk cavities when pulling, pushing and lifting. Ergonomics, *7*:465–474, 1964.
————: Effects on the trunk of erecting pit props at different working heights. Ergonomics, *9*:475–484, 1966.
Davis, P. R., Troup, J. D. G., and Burnard, J. H.: Movements of the thoracic and lumbar spine when lifting: A chrono-cyclophotographic study. J. Anat. Lond., *99*:13–26, 1965.
Donisch, E. W., and Basmajian, J. V.: Electromyography of deep back muscles in man. Am. J. Anat., *133*:25–36, 1972.
Eie, N.: Load capacity of the low back. J. Oslo City Hosp., *16*:73–98, 1966.
Eie, N., and Wehn, P.: Measurements of the intra-abdominal pressure in relation to weight bearing of the lumbosacral spine. J. Oslo City Hosp., *12*:205–217, 1962.
Evans, F. G., and Lissner, H. R.: Biomechanical studies on the lumbar spine and pelvis. J. Bone Joint Surg., *41A*:278–290, 1959.
Farfan, H. F.: *Mechanical Disorders of the Low Back.* Philadelphia, Lea & Febiger, 1973.
————: Muscular mechanism of the lumbar spine and the position of power and efficiency. Orthop. Clin. North Am., *6*:135–144, 1975.
Flint, M. M.: Abdominal muscle involvement during the performance of various forms of sit-up exercise. An electromyographic study. Am. J. Phys. Med., *44*:224–234, 1965.
Floyd, W. F., and Silver, P. H. S.: Electromyographic study of patterns of activity of the anterior abdominal wall muscles in man. J. Anat., *84*:132–145, 1950.
————: The function of the erectores spinae muscles in certain movements and postures in man. J. Physiol., *129*:184–203, 1955.
Friedebold, G.: Die Aktivität normaler Rückenstreckmuskulatur im Elektromyogramm unter verschiedenen Haltungsbedingungen; eine Studie zur Skelettmuskelmechanik. Z. Orthop., *90*:1–18, 1958.
Galante, J. O.: Tensile properties of the human lumbar annulus fibrosus. Acta Orthop. Scand., Suppl. *100*:1–91, 1967.
Gregersen, G. G., and Lucas, D. B.: An *in vivo* study of the axial rotation of the human thoracolumbar spine. J. Bone Joint Surg., *49A*:247–262, 1967.
Hansson, T.: The bone mineral content and biomechanical properties of lumbar vertebrae. An *in vitro* study based on dual photon absorptiometry. Thesis, University of Göteborg, Sweden, 1977.
Hirsch, C., and Lewin, T.: Lumbosacral synovial joints in flexion-extension. Acta Orthop. Scand., *39*:303–311, 1968.
Jayson, M.: *The Lumbar Spine and Back Pain.* New York, Grune and Stratton, 1976.
Jonsson, B.: The functions of individual muscles in the lumbar part of the spinae muscle. Electromyography, *10*:5–21, 1970.
Jørgensen, K.: Back muscle strength and body weight as limiting factors for work in the standing slightly stooped position. Comm. Dan. Nat. Assoc. Infant. Paral., *30*:3–9, 1970.
Joseph, J.: *Man's Posture: Electromyographic Studies.* Springfield, Charles C Thomas, 1960.
Joseph, J., and McColl, I.: Electromyography of muscles of posture: Posterior vertebral muscles in males. J. Physiol., *157*:33–37, 1961.
Kazarian, L. E.: Creep characteristics of the human spinal column. Orthop. Clin. North Am., *6*:3–18, 1975.

Kazarian, L., and Graves, G. A.: Compressive strength characteristics of the human vertebral centrum. Spine, 2:1–14, 1977.

Keegan, J. J.: Alterations of the lumbar curve related to posture and seating. J. Bone Joint Surg., 35A:589–603, 1953.

King, A. I., Prasad, P., and Ewing, C. L.: Mechanism of spinal injury due to caudocephalad acceleration. Orthop. Clin. North Am., 6:19–31, 1975.

Kraus, H.: Effect of lordosis on the stress in the lumbar spine. Clin. Orthop., 117:56–58, 1976.

Kulak, R. F., Schultz, A. B., Belytschko, T., and Galante, J.: Biomechanical characteristics of vertebral motion segments and intervertebral discs. Orthop. Clin. North Am., 6:121–133, 1975.

LeVeau, B. F.: Axes of joint rotation of the lumbar vertebrae during abdominal strengthening exercises. In Biomechanics IV. Edited by R. C. Nelson and C. A. Morehouse. Baltimore, University Park Press, 1974, pp. 361–364.

Lin, H. S., Liu, Y. K., and Adams, K. H.: Mechanical response of the lumbar intervertebral joint under physiological (complex) loading. J. Bone Joint Surg., 60A:41–55, 1978.

Lucas, D. B.: Mechanics of the spine. Bull. Hosp. Joint Dis., 31:115–131, 1970.

Lucas, D. B., and Bresler, B.: Stability of the ligamentous spine. Biomechanics Laboratory, University of California, San Francisco and Berkeley. Technical Report 40. San Francisco, The Laboratory, 1961.

Lumsden, R. M., and Morris, J. M.: An in vivo study of axial rotation and immobilization at the lumbosacral joint. J. Bone Joint Surg., 50A:1591–1602, 1968.

Markolf, K.L.: Deformation of the thoracolumbar intervertebral joints in response to external loads. A biomechanical study using autopsy material. J. Bone Joint Surg., 54A:511–533, 1972.

Miles, M., and Sullivan, W. E.: Lateral bending at the lumbar and lumbosacral joints. Anat. Rec., 139:387–398, 1961.

Moll, J. M. H., and Wright, V.: Normal range of spinal mobility. An objective clinical study. Ann. Rheum. Dis., 30:381–386, 1971.

Morris, J. M., Benner, G., and Lucas, D. B.: An electromyographic study of the intrinsic muscles of the back in man. J. Anat. Lond., 96:509–520, 1962.

Morris, J. M., Lucas, D. B., and Bresler, B.: Role of the trunk in stability of the spine. J. Bone Joint Surg., 43A:327–351, 1961.

Nachemson, A.: Lumbar intradiscal pressure. Acta Orthop. Scand., Suppl. 43:1–104, 1960.

———: The influence of spinal movements on the lumbar intradiscal pressure and on the tensile stresses in the annulus fibrosus. Acta Orthop. Scand., 33:183–207, 1963.

———: Electromyographic studies on the vertebral portion of the psoas muscle. With special reference to its stabilizing function of the lumbar spine. Acta Orthop. Scand., 37:177–190, 1966.

———: The possible importance of the psoas muscle for stabilization of the lumbar spine. Acta Orthop. Scand., 39:47–57, 1968.

———: Towards a better understanding of back pain; a review of the mechanics of the lumbar disc. Rheumatol. Rehabil., 14:129–143, 1975.

Nachemson, A. L.: The lumbar spine: An orthopaedic challenge. Spine, 1:59–71, 1976.

Nachemson, A., and Elfström, G.: Intravital Dynamic Pressure Measurements in Lumbar Discs: A Study of Common Movements, Maneuvers and Exercises. Stockholm, Almqvist & Wiksell, 1970.

Nachemson, A. L., and Evans, J. H.: Some mechanical properties of the third human lumbar interlaminar ligament (ligamentum flavum). J. Biomech., 1:211–220, 1968.

Nachemson, A., and Morris, J. M.: In vivo measurements of intradiscal pressure. Discometry, a method for the determination of pressure in the lower lumbar discs. J. Bone Joint Surg., 46A:1077–1092, 1964.

Naylor, A.: Intervertebral disc prolapse and degeneration. The biochemical and biophysical approach. Spine, 1:108–114, 1976.

Norton, P. L., and Brown, T.: The immobilizing efficiency of back braces. Their effect on the posture and motion of the lumbosacral spine. J. Bone Joint Surg., 39A:111–139, 1957.

Okada, M., Kogi, K., and Ishii, M.: Enduring capacity of the erectores spinae muscles in static work. J. Anthrop. Soc. Nippon, 78:99–110, 1970.

Örtengren, R., and Andersson, G. B. J.: Electromyographic studies of trunk muscles, with special reference to the functional anatomy of the lumbar spine. Spine, 2:44–52, 1977.

Panjabi, M. M., Krag, M. H., White, A. A., and Southwick, W. O.: Effects of preload on load displacement curves of the lumbar spine. Orthop. Clin. North Am., *8*:181–192, 1977.

Park, K. S., and Chaffin, D. B.: A biomechanical evaluation of two methods of manual load lifting. Am. Inst. Indus. Engng. Trans., 6:105–113, 1974.

Partridge, M. J., and Walters, C. E.: Participation of the abdominal muscles in various movements of the trunk in man. An electromyographic study. Phys. Ther. Rev., *39*:791–800, 1959.

Pauly, J. E.: An electromyographic analysis of certain movements and exercises. I. Some deep muscles of the back. Anat. Rec., *155*:223–234, 1966.

Portnoy, H., and Morin, F.: Electromyographic study of postural muscles in various postions and movements. Am. J. Physiol., *186*:122–126, 1956.

Reichmann, S.: Motion of the lumbar articular processes in flexion-extension and lateral flexion of the spine. Acta Morphol. Neerl. Scand., *8*:261–272, 1970/71.

Reichmann, S., Berglund, E., and Lundgren, K.: Das Bewegungszentrum in der Lendenwirbelsäule bei Flexion und Extension. Z. Anat. Entwicklungsgesch., *138*:283–287, 1972.

Rolander, S. D.: Motion of the lumbar spine with special reference to the stabilizing effect of posterior fusion. An experimental study on autopsy specimens. Acta Orthop. Scand., Suppl. *90*:1–144, 1966.

Schultz, A. B., Warwick, D. N., Berkson, M. H., and Nachemson, A. L.: Mechanical properties of human lumbar spine motion segments. Part 1: Responses in flexion, extension, lateral bending, and torsion. J. Biomech. Engng., *101*:46–52, 1979.

Tkaczuk, H.: Tensile properties of human lumbar longitudinal ligaments. Acta Orthop. Scand., Suppl. *115*:1–68, 1968.

Virgin, W. J.: Experimental investigations into the physical properties of the intervertebral disc. J. Bone Joint Surg., *33B*:607–611, 1951.

Waters, R. L., and Morris, J. M.: Effect of spinal supports on the electrical activity of muscles of the trunk. J. Bone Joint Surg., *52A*:51–60, 1970.

Weis, E. B.: Stresses of the lumbosacral junction. Orthop. Clin. North Am., 6:83–91, 1975.

White, A. A.: Analysis of the mechanics of the thoracic spine in man. An experimental study of autopsy specimens. Acta Orthop. Scand., Suppl. *127*:1–105, 1969.

White, A. A., and Hirsch, C.: The significance of the vertebral posterior elements in the mechanics of the thoracic spine. Clin. Orthop., *81*:2–14, 1971.

White, A. A., and Panjabi, M. M.: *Clinical Biomechanics of the Spine*. Philadelphia, J. B. Lippincott Co., 1978.

White, A. A., Southwick, W. O., Panjabi, M. M., and Johnson, R. M.: Practical biomechanics of the spine for the orthopaedic surgeon. American Academy of Orthopaedic Surgeons Instructional Course Lectures, *23*:62–78, 1974.

Glossary of Biomechanical Terms

Abduction—Motion away from the midline

Acceleration—The change in velocity of a body divided by the time over which change occurs

Adduction—Motion toward the midline

Agonistic muscles—Muscles which initiate and carry out motion

Angle of anteversion—The angle of inclination of the femoral neck relative to the femoral shaft in the transverse plane

Angular acceleration—The change in angular velocity of a body divided by the time over which change occurs, usually expressed in radians/sec^2 or degrees/sec^2

Angular deformation—Internal structural change in an angular manner due to shear loading

Anisotropy—The quality whereby a material exhibits unlike mechanical properties when loaded in different directions

Antagonistic muscles—Muscles which oppose the actions of the agonistic muscles

Area moment of inertia—Quantity which takes into account the cross-sectional area and distribution of material around an axis during bending

Asymptotic curve—A graph curve that reaches a certain point after which only one factor need remain constant for one end of the curve to parallel one axis to infinity

Axial rotation—Rotation about an axis

Axis of motion—Line about which all points move in a body in motion

Axis of rotation—Line about which all points in a rotating body describe circles

Bending—A loading mode in which a load is applied to a structure in a manner that causes it to bend about an axis, subjecting the structure to a combination of tension and compression (See *three-point bending* and *four-point bending*)

Bending moment—A quantity at a point in a structure equal to the product of the applied force and the perpendicular distance from the point to the force line, usually measured in newton meters

Bone remodeling—The ability of bone to adapt, by changing its size, shape, and structure, to the mechanical demands placed upon it

Boundary lubrication—A fundamental type of lubrication which depends on the chemical absorption of a monolayer of lubricant molecules between two sliding surfaces (See *fluid film lubrication*)

Brittleness—The quality whereby a material exhibits little deformation before failure

Center of gravity—Equilibrium point of a supported body when all its weight is concentrated (See *center of mass*)

Center of mass—That point at the exact center of an object's mass; often called the center of gravity (See *center of gravity*)

Center of rotation—A point around which circular motion is described

Chondrocyte—A cartilage cell residing in formed matrix

Closed section—A cross section which has a continuous outer surface

Combined loading—Application of two or more loading modes to a structure

Compression—A loading mode in which equal and opposite loads are applied toward the surface of the structure, resulting in shortening and widening

Concentric work—Work produced by a muscle when it is contracting and its length is shortening

Concurrent—Meeting or intersecting in a point

Conservation of momentum—Maintenance of the relationship between velocity and mass

Contact point—The junction point between two joint surfaces

Coplanar—Lying or acting in the same plane

Coronal plane—See *frontal plane*

Coxa valga—Condition in which the angle formed by the axes of the femoral head and shaft is greater than 125 degrees

Coxa vara—Condition in which the angle formed by the axes of the femoral head and shaft is less than 125 degrees

Creep—Progressive deformation of soft tissues due to constant low loading over an extended period of time

Creep displacement—The movement of a structure resulting from creep (See *creep*)

Cross-sectional area—Measure of the area of a piece of material cut at right angles to its longitudinal axis

Deformation rate—The speed at which an applied load deforms a structure (See *speed of loading*)

Degrees of freedom—The number of ways in which a body can move

Density—The mass of matter in a given space

Direction—The path along which motion takes place; with reference to a vector, direction includes line of application and sense

Direction of displacement—The direction of change in position of the contact points of two surfaces

Distraction—The movement of two surfaces away from each other

Dorsiflexion—Bending upwards around an axis

Ductility—The quality whereby a material exhibits extensive deformation before failure

Dynamics—The study of forces acting on a body in motion

Eccentric work—Work produced by a muscle when its length is increasing

Elasticity—Property of a material which allows the material to return to its original shape and size after being deformed

Elastohydrodynamic lubrication—A type of fluid film lubrication which occurs when bearing surfaces are elastic enough for the lubricant pressure to substantially deform the surfaces, resulting in a more substantial and longer lasting film

Equilibrium—State of a body at rest in which the sums of all forces and moments are zero

Extension—The position of the joints of the extremities and back when one stands at rest, or the direction of motion which tends to restore this position; the opposite of flexion

Fatigue curve—A graph plotting the relationship of load and the repetition of the loading which produces fracture

Fatigue fracture—A fracture typically produced by either low repetition of high loads or high repetition of relatively normal loads

Fatigue wear—Removal of material from solid surfaces by mechanical action due to the deformation of the contacting bodies

Flexion—Movement involving the bending of a joint whereby the angle between the bones is diminished; the opposite of extension

Fluid film lubrication—A fundamental type of lubrication wherein a thick film of lubricant causes a relatively large separation of the two bearing surfaces

Force—An action which changes the state of rest or motion of a body to which it is applied

Force couple—Two parallel forces of equal magnitude but opposite direction applied to a structure

Force-elongation curve—See *load-deformation curve*

Four-point bending—A type of bending that takes place when two force couples acting on a structure produce two equal moments (See *bending* and *three-point bending*)

Free body—A structure considered in isolation for the purpose of studying the effect of forces acting on it

Free body diagram—Diagram of an isolated portion of a structure used during free body analysis for the purpose of studying the effect of forces acting on the free body

Friction force—A tangential force opposing motion which acts between two bodies in contact

Frontal plane—The plane which passes through the longitudinal axis of the body

Glycosaminoglycan (GAG)—A long flexible chain of repeating disaccharide units which are the building blocks of proteoglycans

Gravitational force—A force produced by gravitational attraction by the earth on a body

Ground reaction force—A gravitational force produced by the weight of an object against the surface on which it lies

Helicoid—Spiral-like

Horizontal plane—See *transverse plane*

Hydrodynamic lubrication—A form of fluid lubrication wherein relative motion of two bearing surfaces creates a converging wedge of fluid generating a lifting pressure which keeps the surfaces apart

Hydrostatic lubrication—A form of fluid film lubrication in which there is no relative sliding motion of the bearing surfaces; thus this pressure is usually generated by an external pressure supply

Instant center—See *instantaneous center of motion*

Instant center of rotation—See *instantaneous center of motion*

Instant center pathway—A pathway of the instant center for a joint in different positions throughout the range of motion in one plane

Instant center technique—A technique used to describe the relative uniplanar motion of two adjacent segments of a body and the direction of displacement of the contact points between these two segments

Instantaneous center of motion—The immovable point existing at an instant in time created by one segment (link) of a body rotating about an adjacent segment; all other points on the body rotate about this immovable point

Interfacial wear—Removal of material by either adhesion or abrasion due to interaction of the bearing surfaces

Joint lubrication—A design feature of the joint which maintains the continuity of the thin film of synovial fluid between the joint surfaces, minimizing contact and wear of the cartilaginous surfaces

Joint reaction force—The internal reaction force acting at the contact surfaces when a joint in the body is subjected to external loads

Kinematics—The branch of mechanics that deals with motion of a body without reference to force or mass

Kinetics—The branch of mechanics that deals with the motion of a body under the action of given forces

Lever arm—The perpendicular distance from the line of application of a force to the center of motion in a rigid structure, also known as the moment arm of the force

Line of gravity—Action line of the force of gravity

Link—One of two adjacent segments of a body which move about an instantaneous center of motion in a joint

Load-deformation curve—A curve which plots the deformation of a structure when the structure is loaded in a known direction

Load relaxation—Situation where the load decreases with time once the material under loading is deformed to a constant length (See *stress relaxation*)

Loading mode—The manner in which forces are applied to a structure (See *tension, compression, bending, shear, torsion,* and *combined loading*)

Longitudinal axis—A lengthwise line or plane about which a body or system rotates

Longitudinal plane—See *frontal plane*

Mass moment of inertia—The measure of resistance to change in angular velocity, usually expressed in kg m^2 or, equivalently, in Nm sec^2

Matrix—The intercellular substance of a tissue

Modulus—A constant that expresses numerically the degree to which a property is possessed by a material

Modulus of elasticity (Young's modulus)—The ratio of stress to strain at any point in the elastic region of a load-deformation curve, yielding a value for stiffness

Moment arm—See *lever arm*

Neck-shaft angle—The angle of inclination of the femoral neck relative to the femoral shaft in the frontal plane

Moment—A measure of torque; quantity necessary to angularly accelerate a body, usually expressed in newton meters

Neutral axis—The central plane on which the tensile and compressive stresses and strains due to bending equal zero

Normal strain—The amount of deformation in length divided by the structure's original length; the direction of normal strain is perpendicular to the surface of the structure under consideration (See *strain*)

Open section—A cross section of a hollow structure in which the surface of the material is no longer continuous

Open section defect—A large defect which disrupts the continuity of the outer surface of a cylindrical-type structure, roughly equal in its major diameter to the diameter of the cylinder

Permeability—A material parameter representing the frictional resistance of the solid matrix of a porous material to the flow of fluid through it

Perpendicular bisector—A line at right angles to a segment, dividing it into two parts

Physiological cross-sectional area—The amount of muscle fiber in a given cross section of a muscle

Planar—Relating to one plane

Plantar flexion—Bending about the ankle joint in the direction of the sole of the foot

Polar moment of inertia—A quantity which takes into account the cross-sectional area and the distribution of material around a neutral axis in torsional loading

Porosity—The ratio of the volume of interstices of a material to the volume of its mass

Pressure—The surface stress acting perpendicular to a unit area

Prestress—Internal stresses in a material which counteract the stresses that will result from an applied load

Proteoglycan—A macromolecule composed of glycosaminoglycans forming a hydrated gel; one of the primary structural components of cartilage

Range of motion—The range of translation and rotation of a joint for each of its 6 degrees of freedom

Repetitive loading—Repeated application of a load to a structure

Resiliency—The capacity of a strained body to recover its size and shape after deformation.

Rotation—Motion in which all points describe circular arcs about an immovable line or axis

Sagittal plane—The median plane of the body or any plane parallel thereto

Screw-home mechanism—A combination of knee extension and external rotation of the tibia

Shear—A loading mode in which a load is applied parallel to the surface of the structure, causing internal angular deformation

Shear modulus—An innate property of a deformable body that indicates how much resistance the body presents when an attempt is made to shear it, represented by the slope of the shear stress-strain curve; related to the modulus of elasticity

Shear strain—The amount of angular deformation of a structure under shear loading (See *strain*)

Speed of loading—The rate at which load is applied to a structure (See *deformation rate*)

Squeeze film lubrication—A form of fluid lubrication in which the approaching surfaces generate a pressure field in the lubricant as it is forced out of the area of impending contact

Statics—The study of forces acting on a body in equilibrium

Strain—Deformation (lengthening or shortening) of a body divided by its original length

Strain gauge—A device which permits strain to be measured

Strain rate—The speed at which a strain-producing load is applied

Stress—Load per unit area which develops on a plane surface within a structure in response to externally applied loads

Stress concentration effect—The result of stresses concentrating around a small defect in a structure

Stress raiser (stress riser)—Any geometric characteristic (such as a small hole or a sharp internal corner) in a loaded body that causes an abrupt increase in local stress

Stress relaxation—Situation where stress decreases with time once a material under loading is deformed to a constant length (See *load relaxation*)

Stress-strain curve—A curve generated by plotting the stress and the strain during compressive, tensile, or shear loading of a structure

Surface velocity—The surface speed of a body in a given direction

Tangential—Relating to a straight line that is the limiting position of a secant of a curve through a fixed point

Tension—A loading mode in which equal and opposite loads are applied away from the surface of a structure, resulting in lengthening and narrowing

Three-point bending—A type of bending that takes place when three forces act on a structure (See *bending* and *four-point bending)*

Time dependent or viscoelastic (recoverable) material behavior—Material behavior wherein deformation and recovery are influenced by rate and amount of time during loading and unloading

Time independent or elastic (recoverable) material behavior—Material behavior wherein the structure deforms instantaneously when loaded and recovers instantaneously when unloaded

Torsion—A loading mode in which a load is applied to a structure in a manner that causes it to twist about an axis, subjecting the structure to a combination of shear, tension, and compressive loads

Translation—Parallel motion of one surface across another

Transverse—Crosswise; in a horizontal direction

Transverse plane—A plane which extends or lies in a crosswise direction

Ultimate failure point—The point on the load-deformation curve past which complete failure of the structure occurs due to continued loading in the nonelastic region

Uniplanar—See *planar*

Vector—A quantity that has magnitude, sense, line of application, and point of application, commonly represented by a directed line segment

Velocity—The displacement of a body divided by the time over which displacement occurs

Viscosity—The resistance of a fluid to flowing

Wear—The removal of material from solid surfaces by mechanical action (See *interfacial wear* and *fatigue wear)*

Wolff's law—A law which states that bone is laid down where needed and resorbed where not needed

Yield point—The point of the load-deformation curve at which the material begins to yield; if loading continues, permanent deformation results

Index

Page numbers in *italics* refer to figures; page numbers followed by t refer to tables.